T0305955

The Data Preparation Journey

The Data Preparation Journey: Finding Your Way with R introduces the principles of data preparation within in a systematic approach that follows a typical data science or statistical workflow. With that context, readers will work through practical solutions to resolving problems in data using the statistical and data science programming language R. These solutions include examples of complex real-world data, adding greater context and exposing the reader to greater technical challenges. This book focuses on the Import to Tidy to Transform steps. It demonstrates how "Visualise" is an important part of Exploratory Data Analysis, a strategy for identifying potential problems with the data prior to cleaning.

This book is designed for readers with a working knowledge of data manipulation functions in R or other programming languages. It is suitable for academics for whom analyzing data is crucial, businesses who make decisions based on the insights gleaned from collecting data from customer interactions, and public servants who use data to inform policy and program decisions. The principles and practices described within The Data Preparation Journey apply regardless of the context.

Key Features:

- Includes R package containing the code and data sets used in the book
- Comprehensive examples of data preparation from a variety of disciplines
- Defines the key principles of data preparation, from access to publication

Martin Monkman is a Senior Manager at MNP, and a Course Instructor at the University of Victoria Continuing Studies' Business Intelligence and Data Analytics program. Prior to joining MNP, Martin had a long career at BC Stats, the provincial statistics agency in British Columbia, Canada, including a decade with the job title "Provincial Statistician". Martin has Bachelor of Science and Master of Arts degrees in Geography from the University of Victoria, and he has been a member of the Statistical Society of Canada since 2022.

CHAPMAN & HALL/CRC DATA SCIENCE SERIES

Reflecting the interdisciplinary nature of the field, this book series brings together researchers, practitioners, and instructors from statistics, computer science, machine learning, and analytics. The series will publish cutting-edge research, industry applications, and textbooks in data science.

The inclusion of concrete examples, applications, and methods is highly encouraged. The scope of the series includes titles in the areas of machine learning, pattern recognition, predictive analytics, business analytics, Big Data, visualization, programming, software, learning analytics, data wrangling, interactive graphics, and reproducible research.

Recently Published Titles

Data Science for Water Utilities
Data as a Source of Value
Peter Prevos

Practitioner's Guide to Data Science
Hui Lin and Ming Li

Natural Language Processing in the Real World
Text Processing, Analytics, and Classification
Jyotika Singh

Telling Stories with Data
With Applications in R
Rohan Alexander

Big Data Analytics
A Guide to Data Science Practitioners Making the Transition to Big Data
Ulrich Matter

Data Science for Sensory and Consumer Scientists
Thierry Worch, Julien Delarue, Vanessa Rios De Souza and John Ennis

Data Science in Practice
Tom Alby

Introduction to NFL Analytics with R
Bradley J. Congelio

Soccer Analytics: An Introduction Using R
Clive Beggs

Spatial Statistics for Data Science: Theory and Practice with R
Paula Moraga

Research Software Engineering: Research Software Engineering
Matthias Bannert

The Data Preparation Journey: Finding Your Way With R
Martin Hugh Monkman

For more information about this series, please visit: https://www.routledge.com/Chapman--Hall-CRC-Data-Science-Series/book-series/CHDSS

The Data Preparation Journey

Finding Your Way with R

Martin Hugh Monkman

CRC Press is an imprint of the
Taylor & Francis Group, an **informa** business
A CHAPMAN & HALL BOOK

Designed cover image: © Shutterstock, Photo ID 351132254, Photo Contributor jennyt

First edition published 2024
by CRC Press
2385 NW Executive Center Drive, Suite 320, Boca Raton FL 33431

and by CRC Press
4 Park Square, Milton Park, Abingdon, Oxon, OX14 4RN

CRC Press is an imprint of Taylor & Francis Group, LLC

Library of Congress Cataloging-in-Publication Data

Names: Monkman, Martin Hugh, author.
Title: The data preparation journey : finding your way with R / Martin Hugh Monkman.
Description: First edition. | Boca Raton, FL : CRC Press, 2024. |
Series: Chapman & Hall/CRC data science series | Includes bibliographical references and index.
Identifiers: LCCN 2023049317 (print) | LCCN 2023049318 (ebook) |
ISBN 9781032192314 (hardback) | ISBN 9781032189758 (paperback) |
ISBN 9781003258254 (ebook)
Subjects: LCSH: Quantitative research--Data processing. |
R (Computer program language) | Data mining--Computer programs.
Classification: LCC QA76.9.Q36 M66 2024 (print) | LCC QA76.9.Q36 (ebook) |
DDC 006.3/12--dc23/eng/20240229
LC record available at https://lccn.loc.gov/2023049317
LC ebook record available at https://lccn.loc.gov/2023049318

ISBN: 978-1-032-19231-4 (hbk)
ISBN: 978-1-032-18975-8 (pbk)
ISBN: 978-1-003-25825-4 (ebk)

DOI: 10.1201/9781003258254

Typeset in Latin Modern font
by KnowledgeWorks Global Ltd.

Publisher's note: This book has been prepared from camera-ready copy provided by the authors.

For Jamie.

Contents

Preface

It is routinely noted that the Pareto principle[1] applies to data science—80% of one's time is spent on data collection and preparation, and the remaining 20% on the "fun stuff" like modelling, data visualization, and communication.

There is no shortage of material—textbooks, journal articles, blog posts, online courses, podcasts, etc.—about the 20%. That's not to say that there is no material for the other 80%. But it is scattered, found across technique-specific articles and domain-specific books, along with Stack Overflow questions and miscellaneous blog posts. This book serves as a travel guide: an introduction and wayfinder through some of the scattered resources for readers seeking to understand the core elements of data preparation. It is hoped that, like a lighthouse, it will both guide you in the right direction and keep you from running aground.

The book will introduce the principles of data preparation, framed in a systematic approach that follows a typical data science or statistical workflow. With that context, readers will then work through practical solutions to resolving problems in data using the statistical and data science programming language R. These solutions will include examples of complex real-world data.

In *Exploratory Data Analysis*, Tukey writes "the analyst of data needs both tools and understanding. The purpose of this book is to provide some of each." (Tukey, 1977, p.1) It is my modest hope that this book also provides you both tools and understanding.

You, the reader

You might be an academic, working in the physical sciences, social sciences, or humanities, who is (or will be) analyzing data as part of your research. You might be working in a business setting, where important decisions are being made based on the insights you draw from the data collected as part of interactions with customers. As a public servant, you might be creating the

[1] https://en.wikipedia.org/wiki/Pareto_principle

evidence a government or other public agency is using to inform policy and program decisions. The principles and practices described in this book will apply no matter the context.

It is assumed that the reader of this book will have a working knowledge of the fundamental data manipulation functions in R (whether base or tidyverse or packages beyond those) or another programming language that supports that work. If you can filter for specific values in the variables and select the columns you want, know the difference between a character string and a numeric value (`"1"` or `1`), and can create a new variable as the result of a manipulation of others, then we're on our way.

This book leans heavily on R Markdown, particularly when it comes to describing documentation and the packages of the tidyverse. Familiarity with both will be very helpful.

If you don't possess that knowledge yet, I would recommend that you work through *R for Data Science (2nd edition)* by Hadley Wickham, Mine Çetinkaya-Rundel, and Garrett Grolemund (Wickham et al., 2023b). This book, freely available at 4ds.hadley.nz[2], will give you a running start.

While some of the topics covered here may be similar to what you'll find in *R for Data Science* and other introductory books and similar resources, it is hoped that the examples in this book add more context and expose you to greater technical challenges.

Outline

The first three chapters of this book provide some foundations, elements of the data preparation process that will help guide our thinking and our work, including data documentation (or recordkeeping).

Chapters 4 through 10 cover importing data from a variety of sources that are commonly encountered, including plain-text, Excel, statistical software formats, PDF files, internet sources, and databases.

Chapters 11 and 12 tackle finding problems in our data, and then dealing with those problems.

Finally Chapter 13 presents a short summary and poses the question, "Where to from here?"

[2]https://r4ds.hadley.nz/

Acknowledgements

I would like to acknowledge everyone who has contributed to the books, articles, blog posts, and R packages cited within. As well, thanks to my current colleagues at MNP, my former colleagues at BC Stats, and my colleagues and students at the University of Victoria's Business Intelligence & Data Analytics program. The enthusiasm of this community of people—some I know well and others around the world I have never met—has helped sustain my own interest, and without that interest I wouldn't have written this book.

To the people at Taylor & Francis who have supported my efforts in writing this book. In particular, thanks are due to David Grubbs, Curtis Hill, Robin Lloyd-Starkes, and Lee Baldwin.

Particular thanks to Julie Hawkins and Emily Riederer, both of whom provided valuable feedback on early drafts, and through their critiques made this book much better than it would have otherwise been.

And finally, to my family, thanks for your unfailing enthusiasm for this project.

Land Acknowledgement

I acknowledge with respect the Lekwungen-speaking peoples on whose traditional territory I live and work, and the Songhees, Esquimalt, and W̱SÁNEĆ peoples whose historical relationships with the land continue to this day.

Some important details

License

[3]https://github.com/MonkmanMH

[4]https://creativecommons.org/licenses/by-nc-nd/4.0/

Data in this book

This book draws on three sources of data:

- R packages, such as {palmerpenguins} and {Lahman}, which bundle data ready for use.
- Open source data that is freely available on the web.
- Mock or synthetic data that was created specifically for this book.

The various data files in the second and third groups are bundled in the R package {dpjr} (Monkman, 2024).

To download and install the {dpjr} package, you will need the {remotes} package:

```
# download and install "remotes"
install.packages("remotes")
# download and install "dpjr"
remotes::install_github("monkmanmh/dpjr")
```

Once the package is downloaded, the function `dpjr::dpjr_data(<filename>)` can be used to generate the path to the data file, independent of the specific location on the user's computer.

For example, to read the CSV file "mpg.csv":

```
df_mtcars <- read.csv(dpjr::dpjr_data("mpg.csv"))
```

The {dpjr} package website is here: https://monkmanmh.github.io/dpjr/

The data files used in the {dpjr} package are covered by various open licenses; details can be found at the "Data licenses" page at the package website.

Source code

The source code for this ebook can be found at this github repository: https://github.com/MonkmanMH/data_preparation_journey

This book is written in **Markdown**, using the {bookdown} package (Xie, 2021), and published to the web at bookdown.org[5].

[5]https://bookdown.org/

```
install.packages("bookdown")
# or the development version
# devtools::install_github("rstudio/bookdown")
```

Cover image

The cover image is a wayfinder close to my home: Fisgard Lighthouse, marking the entrance to Esquimalt Harbour in Victoria, British Columbia, Canada. (Location: 48.4307, -123.4477)

The cover photo is by jennyt, and used with permission.

About the Author

My professional bio reads "Martin Monkman has three decades of expertise supporting evidence-based policy and business management decisions by gathering data from existing sources and through original collection (including surveys), then using data science and statistical methods to generate understanding, and communicating that information through reports and visualization."

What does this really mean? I have spent a long time finding data—often very messy data—and then using different analytic techniques to provide insights through statistical and data science methods, which in turn help organizations make better decisions. I spent many years working as part of BC Stats, the provincial statistics bureau in British Columbia, Canada, including a decade with the job title "Provincial Statistician". In that context, the questions answered by the data analysis were often associated with public policy and the operations of public agencies (including human resources research). I now do similar work for public and private sector clients as part of the Consulting Services team at MNP, one of Canada's leading professional service firms. These experiences, including some of the in-real-life data that I have encountered along the way, have influenced what is in this book.

I also teach courses in the Business Intelligence and Data Analytics program at the University of Victoria's Continuing Studies department. This experience has also informed the content of this book—some of the lessons in those courses have found their way into this book.

I hold Bachelor of Science and Master of Arts degrees in Geography from the University of Victoria. I am a Carpentries Instructor, and also a member of the Statistical Society of Canada (SSC) and the Society for American Baseball Research (SABR).

About the Author

1

Introduction

1.1 The origin of data

Data is proliferating, collected as part of scientific research (whether measuring sub-atomic particles or galaxies in the early universe, plant growth or the migration of animals, people's socio-economic characteristics or where they live and work in cities), and by the tools that are used to conduct the business of the world.

Today, data about *you* might be collected when you:

- Buy a cup of coffee at a café. This generates at least three types of data: about the type of drink you ordered, about your coffee habits through the rewards program, and about the financial transaction—which are used by the store to measure sales and by you to manage your personal budget.
- Use a streaming service to watch a movie or listen to some music. The service provider builds a profile about your preferences, which it can then use to make recommendations as to what you might also enjoy. The service can also use the aggregated data from all subscribers to evaluate the success of the content on their platform, and pay the creators of the content.
- Renew your driver's license. Information about you such as date of birth, height, weight, and current address is recorded, which you can then use as a piece of identification to authenticate who you are. The licensing agency can also use the data in aggregate to estimate how many people will be renewing their licenses each month, and often share the data (securely, of course!) with other agencies. In British Columbia, the agency that licenses drivers shares elements of the database, such as your current address, with the provincial electoral agency who uses that information to maintain an up-to-date list of eligible voters in the province. (Elections BC, 2018, page 70)
- Travel to your place of work or study, captured through the movement of your cell phone. Aggregated location and movement can be used for a variety of purposes, including use by GPS systems to indicate areas of

road congestion, or to let you know when the next bus will arrive at your stop.

- Respond to a survey, perhaps asking for your opinions about the platforms of various political parties and the people who represent those parties.

1.2 Analyzing data: the data science process

A lot has been written about the benefits of organizations being "data-driven", when good data analysis leads to better decisions. There is no limit to the contexts where data analysis is being applied. In the private sector, businesses can compete on analytics (Davenport and Harris, 2007), including retailers (Colson et al., 2017) and supply chain management (Ashton, 2018). It can also be used in the public sector: data analysis can inform public health strategies during a pandemic (Polonsky et al., 2019).

The process of going from raw data, collected during processes like those described above, to support decision making is often called data science. Data science combines the academic disciplines of statistics, computer programming, and the subject matter (be it astronomy, psychology, economics, or business). (Conway, 2010)

The typical data science process has been described in this model (Wickham et al., 2023b):

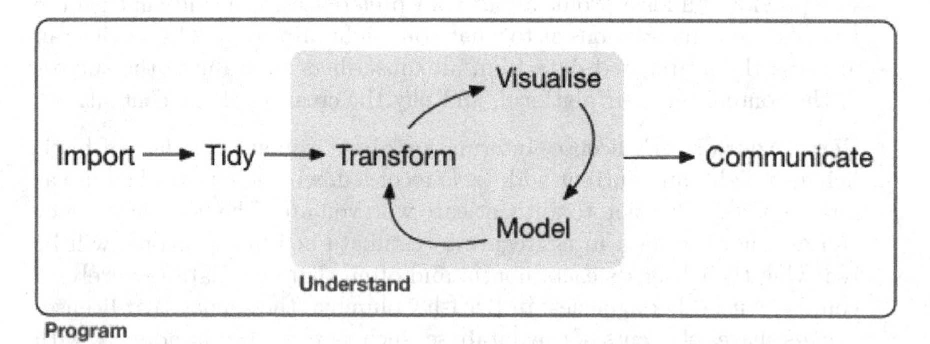

FIGURE 1.1 *The data science process.*

The generalist data scientist has a hand in all the steps through this workflow (Monkman, 2019); in this book we will focus on the Import to Tidy to Transform steps. We will also see how "Visualise" is an important part of Exploratory Data Analysis, a strategy for identifying potential problems with the data prior to cleaning.

1.3 Data in the wild

Most data science and statistics textbooks use example data sets that are nicely formatted and easy to import. Anscombe's Quartet (Anscombe, 1973) is a famous example of a small data set that was designed to produce specific statistical results; it is included in the base version of R (R Core Team, 2021). Other data sets might be excerpted to use as teaching examples (Bryan, 2017), chosen for their statistical properties (Horst et al., 2022), compiled as examples to demonstrate a methodology or technique (Kuhn, 2020), or used to clean, validate, and aggregate still-larger data collections (Friendly et al., 2020).

But in real life the data we find ourselves working with is, more often than not, entirely different from those text book examples.

> *Classroom data are like teddy bears; real data are like a grizzly with salmon blood dripping out of its mouth*[1].

- Data can be stored in different formats–spreadsheets, databases, software-specific file formats, websites, and PDF files. We need to have tools to access the data, no matter the file format.
- The data might be stored in an untidy layout (Wickham, 2014), or inconsistently structured.
- We might receive data from multiple sources, so we end up with two files that have the same value but are coded in different ways. For example, my home province in Canada can be represented as "British Columbia", "B.C.", or "BC". All three are accurate, but these variations can cause problems if they are inconsistent across the data sources (or even more problematically, in the same variable). And don't get me started on date formats—we'll save that for later.
- Data might have as-of-yet undetected errors. It might be incomplete, inconsistent, or it might have data entry mistakes and typos.
- It might have bias...and there are many different types of bias.

[1]quote attributed to Jenny Bryan, "Teach Data Science and They Will Come," Joint Statistical Meetings, 2015

- The data may come from an administrative source, and collected for a purpose other than research. In these cases, you may need to transform that data to fit the requirements of the research question.

This book will introduce steps necessary to ensure that you can assemble your data in a way that allows you to undertake the rest of the data science process, including sharing the data with other data analysts who might seek to use the data in further analysis.

1.4 Data collection

Before we can start our data science workflow—before we get to the "Import" step—someone has to collect the data. This book *won't* cover the process of collecting data, a topic both broad and deep. Included in that would be determining the research question (or hypothesis) or business need that the data will support, designing the experiment or the business process to be captured (both could include a survey questionnaire), and undertaking the experiment (including fieldwork) or implementing the business data collection. The details of the data collection process are often domain-specific, so I encourage you to seek out reference materials that are tailored for your chosen subject area.

2

Foundations

In this chapter:

- The data preparation process
- The principles of data quality, and how dirty data is part of data quality
- Understanding what makes data "clean" or "dirty", and why context matters
- Understanding tidy data principles

2.1 The data preparation process

As we get started on the path of preparing our data, we should think first about where we want to go.

Ellis and Leek's article "How to Share Data for Collaboration" (Ellis and Leek, 2017) is aimed at scientists ("data generators") who have collected the data and are preparing it for further analysis in an academic environment—the authors speak about "preparing data for a statistician."

While Ellis and Leek have a specific context in mind, the principles in their paper have broad applicability, and the practices are essential in any environment, including a business, government, or non-profit organization. And in those contexts, the roles and responsibilities and the division of labour in the workflow, are often different than in an academic environment. The data collection and storage might be to support a business need, and the use of the data for business intelligence is a secondary benefit. As a result, there is often a middle person who isn't involved in the collection of the data but who does the preparation of the data for the modelling, visualization, and reporting. That same person might also be responsible for the modelling, visualization, and reporting.

At the end of the data preparation phase, before visualization and modelling can start, there should be:

1. The raw data.

2. A tidy dataset (Wickham, 2014).

3. A code book describing each variable and its values in the tidy dataset.

4. An explicit and exact recipe used by the researcher to go from 1 to 2 to 3.

Over the next few chapters in this book, we will look at examples of creating a tidy data set (including cleaning the data), creating the code book, and documenting our steps along the way, thereby creating an "explicit and exact recipe."

Whether we are preparing data as part of our own analytic project or to make it available for sharing, these should be our goals.

2.1.1 Elements of an iterative process

This list below are things that make up the process you will go through in preparing your data for analysis and modelling. These are not "steps", in that data preparation is an iterative process, it's not linear or even necessarily sequential.

1. Save a copy of the original data. Do not make any edits, and consider making the file "read-only".
2. Start a new blank "readme" file, and in that file record a few key points, including the project's research objective and information about the data file's origin. You can also draft an outline of the steps that you think you're going to take in the preparation process, as the first contribution to a literate programming approach.
3. Import the data into your **R** environment.
4. Explore and validate the data, assessing the structure, looking for missing values and other "dirty" elements.
5. Clean the data. (And then validate to make sure your cleaning has been successful.)

This whole process is iterative (you will be adding to the readme file at every step), and the individual steps themselves are iterative. For example, your first code to import the data may be revisited, as you evaluate and assign variable types through the various arguments in the read function. Or your first cleaning code may address one dirty element, only to expose a second that you hadn't previously identified.

In addition, you will be using your data wrangling, manipulation, and visualization skills at various points along the way.

2.2 Data quality

Part of the data preparation process is to ask, "What is the data's *quality*?"

One resource that is useful to frame your thinking on this is Statistics Canada's *Quality Guidelines* (Statistics Canada, 2019). Like other official statistical agencies around the world, Statistics Canada's reputation is staked on making high quality data available to the widest possible audience. They have identified six elements of data quality.

- Relevance: The degree to which the data meets the user's needs and relates to the issues that the user cares about.
- Timeliness and punctuality: The delay between the information reference period and the date when the data becomes available.
- Accuracy and reliability: Accuracy "is the degree to which the information correctly describes the phenomena it was designed to measure."
- Accessibility and clarity: "The ease with which users can learn that the information (including metadata) exists, find it, view it and import it into their own work environment."
- Interpretability: "The availability of supplementary information and metadata needed to interpret and use statistical information appropriately."
- Coherence and comparability: "The degree to which it can be reliably combined and compared with other statistical information within a broad analytical framework over time."

In our data preparation process, we want to ensure our own work is accessible and interpretable; this is the motivation behind documentation.

Another typology of "data quality" is found in (Wang et al., 1995). High quality data is:

- Accessible
- Interpretable
- Useful
- Believable

"Dirty data" is data that falls short on the *believable* dimension, in particular, evaluating whether the data are complete, consistent, and accurate. It is important to note that these categories are not mutually exclusive; a variable might be simultaneously inconsistent and inaccurate.

Complete

For our purposes, "complete" means whether any values in each record are missing (internally complete).

"Complete" does not mean that every possible member of the population is represented in the data. With a well-designed sample, it is possible that a sub-set of the population can provide an accurate set of measures about the population. Furthermore, it is possible to determine whether the records are an accurate representation of the whole.

Consistent

We will consider a measure to be "consistent" if the same value is reported in the same way.

Some examples of consistency:

- Units are consistent. One example is temperature, and ensuring that the values are consistently reporting in degrees Celsius, not mixing Fahrenheit and Kelvin. Another might be in a survey or form with an international audience, where `salary` might be completed by respondents in the values of their local currency.
- Spelling is consistent. My home province in Canada is "British Columbia", but is often abbreviated to "B.C." or "BC". Or consider the 57 different ways that "Philadelphia" was spelled in the US Paycheck Protection Program (PPP) applications. (Au, 2020a)
- In some cases, the mode of data collection can introduce differences in the value recorded. This can apply to everything from variability of scientific instruments such as air quality sensors (Khreis et al., 2008) to how people respond to surveys conducted in different media (Abeysundera, 2015) (Holbrook et al., 2003) (St-Pierre and Béland, 2004).

Accurate

When we say "accurate", we mean that the value recorded in our data is the value measured.

Some examples of inaccurate data are:

- The use of a default (or sentinel) value, inserted as a placeholder for unknown or missing values in place of a missing or "unknown" value. This is sometimes part of the data collection or database software, leading to values that should have been "unknown" being entered as the default. This can be the consequence of a data entry validation process that requires an entry (that is, data entry cannot continue until the field has an entry) or an entry in a particular format, such as a date. In these cases, the sentinel value can be entered instead of an explicit "NA".

> Dr Davis Lawrence, director of safety-literature database the SafeLit Foundation...tells me that 'in most US states the quality of police crash reports is at best poor for use as a research tool. ... Data-quality checks were rare and when quality was evaluated it was found wanting. For example, in Louisiana for most crashes in the 1980s most of the occupants were males who were born on January 1st, 1950. Almost all of the vehicles involved in crashes were the 1960 model year.' Except they weren't. These were just the default settings. (Perez, 2019, p.190)

- A data entry error. Perhaps the most common are typographical errors, where the wrong value is entered. Another type is what we might call a "variable transposition error" (where the value is transposed one column over, something that happens all too often with address records, where the city name might end up in the state/province column).
- Values are transformed automagically by software. Microsoft's Excel spreadsheet program is the most famous of these, converting many non-date values into date format, and assigning new values to store the data. For example, entering the character string "SEPT1" gets converted to September 1st of the current year. There is documented evidence that this software behaviour has caused errors in gene research. (Zeeberg et al., 2004), (Abeysooriya et al., 2021)
- Dates may be stored as a numeric value (or "serial number") representing the number of days elapsed from a fixed starting point—a starting point that varies by operating system. For example, Excel for Windows uses January 1, 1900 as the first day, while earlier version Macintosh computers, and by extension Excel for Macintosh, used January 1, 1904 as the start date. Thus the same date would be stored as different values[1]. (https://docs.microsoft .com/en-us/office/troubleshoot/excel/1900-and-1904-date-system)

[1] "Differences between the 1900 and the 1904 date system in Excel"

- Contradiction. For example, imagine that a single individual is entered into two databases, but when we compare them we see that the date of birth differs, perhaps due to a non-ISO8601 entry: 07-08-79 and 08-07-79 both have the same digits but one could be mm-dd-yy and the other dd-mm-yy...we just don't know which is the correct one. Or is one a typo?
- Cultural ignorance (for want of a better term).

> "Prawo Jazdy" was a supposed Polish national who was listed by the Garda Síochána in a police criminal database as having committed more than 50 traffic violations in Ireland. A 2007 memorandum stated that an investigation revealed prawo jazdy [pra.v jaz.d] to be Polish for 'driving licence', with the error arising due to officers mistaking the phrase, printed on Polish driving licences, to be a personal name while issuing traffic tickets. (Wikipedia contributors, 2021)

2.2.1 Tidy data

When we are cleaning our data, we should also consider the structure. The goal should be a tidy structure, one that meets the following three principles or rules:

- Each variable must have its own column.
- Each observation must have its own row.
- Each value must have its own cell. (Wickham, 2014)

In many instances, achieving a tidy structure may require reshaping the data, using the {tidyr}(Wickham, 2021b) packages's `pivot_()` functions. In others, the data may have multiple variables in each cell which need to be split (such as addresses) into a separate column for each variable.

Other common structural problems that fall outside the four broad data quality categories include:

1. Column headers are values instead of variable names. A frequently-encountered example are units of time, as in the example below. In the table below, the "values" are the years. Tables structured in this

way can only display the values of one variable; in this case, it is the population of the country.

country	1952	2007
Canada	14,785,584	33,390,141
United States	157,553,000	301,139,947

A preferable structure would be to have a variable "year" and another "population". This tidy structure also permits other variables measured during those years to be incorporated, rather than displayed in a separate table.

country	year	population	life expectancy
Canada	1952	14,785,584	68.75
Canada	2007	33,390,141	80.65
United States	1952	157,553,000	68.44
United States	2007	301,139,947	78.24

2. Variable names that are duplicated (this is possible in some data storage formats, including plain-text and Excel files).
3. Inappropriate data types, such as numbers or logical values stored as character strings.
4. Multiple data types in a column. For example, some records are numbers and some are character strings. The values themselves could be entirely correct, but if some number entry includes commas as thousands separators, the variable will be read and stored as a character type.

2.3 Cleaning the data

If the data fails to meet our standards or quality, we need to *clean the data.* Which doesn't sound like a lot of fun. Didn't we want to be data scientists or business intelligence experts or academic scientists uncovering the insights hiding in the data? Don't we want to be *doing analysis*?

There is a strong argument to be made that the process of cleaning data is a fundamental part of the analytic process, or the corresponding statement that any analytic process requires data cleaning.

Randy Au has written “The act of cleaning data imposes values/judgments/interpretations upon data intended to allow downstream analysis algorithms to function and give results. That’s exactly the same as doing data analysis. In fact, “cleaning” is just a spectrum of reusable data transformations on the path towards doing a full data analysis.” (Au, 2020a)

At this point we should ask what are we doing when we say we are “cleaning the data”? And how can we confirm that it is “cleaned” in the way that we have defined?

The first challenge: How do we find the things that are problematic with our data?

The second challenge: What can and should we do about them?

We will see some of these challenges and solutions in the next few chapters, as part of the data import process. We will also return to explicitly address these challenges in Chapters 11 and 12.

3

Data documentation

In this chapter:

- Why we want to create documentation
- What elements to record
- Project information, including data dictionaries
- Code documentation through literate programming

3.1 Introduction

The documentation of the data sources, methods, and workflow is vital for reproducibility. There are two places where we need to embrace documentation: the first is within the code scripts, and the second is what we might consider the meta-information about our projects.

"If you're not thinking about keeping track of things, you won't keep track of things." —Karl Broman

Documentation is also essential if you intend to share your data (White et al., 2013). The documentation we create provides metadata (which we will see more of in the data dictionary section below).

3.2 Documentation and recordkeeping

The reasons for documenting our work are similar to the reasons we would save and output our cleaned data file:

- To make the project available to a collaborator.

- To provide a record for others who might use the data in the future. Your project might repeat a year or more in the future, after you have moved on to a new role. Leaving good documentation is good karma, or a "pay it forward in advance"—imagine how good that future user is going to feel when they find your clear explanations!

- To create a save point in our workflow, so that we can continue our work later but without having to go through all the steps to remember everything we've done.

As well, things change (the data, the software and hardware we use, the motivations for our work) and capturing the details about those things at the point we implement them helps us make any necessary changes that might be required at a later date.

There doesn't seem to be a single-best how-to instruction on "documentation" in the context of data analytics or data science. What advice exists tends to be prescriptive and at a high level, and not overly descriptive of what makes good documentation. The term appears in some of the textbooks and manuals on how to do data science, and there are some journal articles that address documentation in the context of reproducible research. In the blog post "Let's Get Intentional About Documentation," Randy Au presents the case that a better term might be "recordkeeping".

"Recordkeeping is keeping records, notes, and artifacts around to **help in creating documentation in the future.** It's done by the people involved in doing the actual work saving things and making clarifications that only they really understand, with no expectation that someone else needs to understand everything. The material collected is primarily intended for use to write the actual documentation. Think of it as preserving raw data for future use." (Au, 2020b)

I like the way this is framed. In this view, "documentation" becomes a fully-realized user manual, suitable for complex and recurring projects with many different contributors, and that might be considered mission critical. A lot of data science work doesn't meet those criteria, so what is needed isn't "documentation" but the notes that someone could use to build the documentation.

If it's a one-off project, you can probably have minimal (but not zero!) recordkeeping. But a project that repeats on a regular schedule (whether daily, monthly, or annually) should have more robust recordkeeping, and potentially some documentation.

And it's important to think about this at the beginning of your work, perhaps even *before* the beginning. Taking this approach, recordkeeping gets built into the process and your daily workflow, and becomes part of the work of data analysis. For example, you could record the source of your data as you're downloading it, or make comments about why you chose a particular function as soon as you incorporate the function into your script. These habits will not only capture what you're thinking at the time, but will also encourage you to think about the "why" of your decisions—and perhaps lead you to a different choice.

The idea of reproducibility is a fundamental element of good scientific practice, and should be extended to include the analysis of the data. "The standard of reproducibility calls for the data and the computer code used to analyze the data be made available to others." (Peng, 2011, p.1226)

And while it's important in an academic environment, this necessity for reproducibility extends to business and government settings as well. In any situation where the data analysis is supporting decisions, it is imperative that the analyst (i.e. you) be able to explain their methods, and quickly and effectively respond to changes in the data (you might receive an update to the data with more records or revisions to the existing records).

The benefits of a reproducible data science process are summarized here:

- Saves time,
- Produces better science,
- Creates more trusted research,
- Reduces the risk of errors, and
- Encourages collaboration. (Smith, 2017)

All of the details involved with creating a reproducible workflow are outside the scope of this book; here we will concern ourselves with documentation.

3.2.1 Elements of effective recordkeeping

There are three things to capture in your recordkeeping:

1. The why and how of the decisions made.
2. How things work together.
3. How to make changes.

3.3 Code documentation

"Document and comment your code for yourself as if you will need to understand it in 6 months." — attributed to Jenny Bryan

In the journal article "Good enough practices in scientific computing", the authors write "Good documentation is a key factor in software adoption, but in practice, people won't write comprehensive documentation until they have collaborators who will use it. They will, however, quickly see the point of a brief explanatory comment at the start of each script, so we have recommended that as a first step." (Wilson et al., 2017)

3.3.1 Literate programming

Computer scientist Donald Knuth introduced the idea of "literate programming" in 1984. In this approach, a natural language explanation takes a central role, and the code chunks for each step follow that explanation. The idea is to "code your documentation," instead of "document your code."

Knuth wrote "Let us change our traditional attitude to the construction of programs: Instead of imagining that our main task is to instruct a computer what to do, let us concentrate rather on explaining to human beings what we want a computer to do." (Knuth, 1992, p.99)

In the context of a data science project, this is enabled in R Markdown (or Jupyter notebooks), where text descriptions of the what and why of a code chunk can precede that chunk. So rather than existing as a separate "documentation" document, the record of the programmer's thoughts are captured in the file that runs the code.

These two quotes capture the key concepts of code documentation:

- "Document interfaces and reasons, not implementations." (Wilson et al., 2014)
- "Comments should explain the why, not the what." (Wickham, 2015, p.68)

Using plain language in these notes also encourages you to think a second time about what you have just coded. Explaining the code to an imagined future reader (or your rubber duck) may help you to see a better way to write your code.

3.3.2 The instruction list

The instruction list is descriptive text that describes what chunks of code are doing.

In "How to Share Data for Collaboration," Ellis and Leek (Ellis and Leek, 2017) suggest something that is essential is a high-level outline of what the code is doing. Here's a rephrase of their first two steps:

- Step 1: Take the raw data file, run summarize code with parameters $a = 1$, $b = 2$
- Step 2: Run the code separately for each sample

To then apply Knuth's "Literate Programming" ideas, each step would then be followed by the code to carry out that step. The R Markdown format is ideal for this sort of work; the text description would be followed by a chunk of R code, followed by another text description/R chunk pair, and so on.

In practice, you may find yourself taking a step-by-step approach:

1. Create an outline of the basic plan,
2. Within that plan add the steps that you will take in your analysis,
3. Add some text detail for each step,
4. Write the code.

3.3.3 An example of documented code

What do these principles look like in practice? This example shows what each step might look like.

We start with the Outline:

- Objective: Find relative sizes of the 10 largest economies in the world.
- Data: The {gapminder} package will be used, using the most recent year of data (2007).
- Calculation: Per capita GDP * population = total country GDP.
- Table: Create table with top 10 countries.

With that outline, we can proceed into the "Data" step

```
# DATA

gapminder_gdp <- gapminder::gapminder |>
  # filter for most recent year
  # use `slice_max() function instead of hard-coding value
  # - make sure that it is the most recent data
  # note: this is a shorthand for `filter(year == max(year))`
  slice_max(year)
```

The "Calculation" step:

```
# CALCULATION

gapminder_gdp <- gapminder_gdp |>
  # calculate total GDP
  mutate(gdpCountry = gdpPercap * pop) |>
  # divide by 10^12 to turn into trillions
  mutate(gdpCountry = gdpCountry/(10^12)) |>
  # select relevant variables
  select(country, gdpCountry)

head(gapminder_gdp)
```

```
## # A tibble: 6 x 2
##   country     gdpCountry
##   <fct>            <dbl>
## 1 Afghanistan     0.0311
## 2 Albania         0.0214
## 3 Algeria         0.207
## 4 Angola          0.0596
## 5 Argentina       0.515
## 6 Australia       0.704
```

Finally, creating the table and formatting it with the {gt} package (Iannone et al., 2023)

```
gapminder_gdp |>
  # use `slice_max()` with `n =` argument to get 10 largest economies
  # `slice_max()` has the added bonus of sorting in descending order
  slice_max(gdpCountry, n = 10) |>
  # create table with {gt} package
  gt::gt() |>
    # add table formatting
    fmt_number() |>
    # add table text
    tab_header(
      title = "Size of 10 largest national economies",
      subtitle = "Total GDP, 2007 (trillions of US$)"
    ) |>
    tab_source_note(
    source_note = "Source: {gapminder} package"
    )
```

Size of 10 largest national economies
Total GDP, 2007 (trillions of US$)

country	gdpCountry
United States	12.93
China	6.54
Japan	4.04
India	2.72
Germany	2.65
United Kingdom	2.02
France	1.86
Brazil	1.72
Italy	1.66
Mexico	1.30

Source: {gapminder} package

```
# note: other table formatting to consider:
# - change variable names
# - add growth from 1957?
```

The first of Benjamin D. Lee's "Ten simple rules for documenting scientific software" (Lee, 2018) is "Write comments as you code."

3.4 Project documentation

Part of the higher-level documentation is the structure of the files in your project.

For those working in an academic setting, Marwick, Boettiger, and Mullen (Marwick et al., 2018) have defined three generic principles for what should be included in research compendia. These principles can also be adapted to other settings:

- Organize the files according to the prevailing standards.
- Maintain a clear separation of data, method, and output.
- Specify the computational environment that was used.

This information is also referred to as a project's "metadata" (White et al., 2013).

A more detailed list of the things to capture in documentation includes:

- A brief description of the project, analysis, and research objective.
- A map of the file folder structure.
- Detailed information about the data source
  - Name of the organization that collected the data.
  - Name and contact information of the individual at that organization.
  - Date of download or extract.
  - Web links.
  - Original file names.
  - Links to additional documentation about the data, such as the data dictionary.

- A high-level statement about *how* the analysis was done
  - For example: "This analysis used the daily data from 2010 through 2019 to calculate average per month volumes, which were then used in an ARIMA model to develop a 6 month forecast."
- Details about the analysis and code:
  - Things that informed decisions ("I chose this statistical test because..." or "The variable date of birth was transformed into 'Age on July 1, 2020' because...")
  - The instruction list—descriptive text that describes what files or chunks of code are doing, but not how they are doing it.
  - The data dictionary or code book.
- Details about the software and packages used:
  - Versions and (if available) release dates.

3.4.1 README files

> Benjamin D. Lee's fourth rule in "Ten simple rules for documenting scientific software" (Lee, 2018) is "Include a README file with basic information."

Having a README file in the root folder of your project is a good place to capture a great deal of the important general information about the project. And having another README in the data folder wouldn't hurt either.

Another source that describes a README for data files is the "Guide to writing"readme" style metadata" written by the Research Data Management Service Group at Cornell University. (Research Data Management Service Group, 2021)

The SFBrigade, a San Francisco volunteer civic tech group, has created a data science specific README outline for their projects, hosted at github[1].

The basic minimum README structure recommended is:

[1]Project-README-template.md(https://github.com/sfbrigade/data-science-wg/blob/master/dswg_project_resources/Project-README-template.md)

- Project Intro/Objective
- Project Description
  - A missing component is "Data source".
- Team (with contact information)

For an analytic project that has been completed, you might also want to include the high-level findings. This might be something akin to an abstract in an academic journal article or an executive summary in a business or government report.

Another source of tips is "Art of README"[2]. Although it's aimed at people writing computer programs and modules, the lessons apply to data science projects as well. The purpose of a README is to:

1. Tell them what it is (with context).
2. Show them what it looks like in action.
3. Show them how they use it.
4. Tell them any other relevant details.

An example is the R package {palmerpenguins} which has a comprehensive README on the github page.[3]

3.4.2 Capturing the directory structure

Your documentation (perhaps as part of the README file) can include a diagram that shows the directory structure. The package {fs} (Hester et al., 2020) has a function `dir_tree()` that will create a dendogram of the file folder structure and files in your project folder.

```
library(fs)
```

The default is to show all of the sub-folders and files below the current directory.

```
fs::dir_tree(path = "project_template")
```

```
## project_template
```

[2]https://github.com/noffle/art-of-readme

[3]https://github.com/allisonhorst/palmerpenguins/blob/master/README.md

```
## +-- data
## |   \-- readme.md
## +-- data-raw
## |   +-- readme.md
## |   \-- source_data.xlsx
## +-- figs
## |   \-- readme.md
## +-- output
## |   \-- readme.md
## +-- readme.md
## \-- scripts
##     \-- readme.md
```

The `recurse =` argument specifies how deep into the directory structure the diagram will be drawn. Specifying `recurse = FALSE` limits what is shown to the files and sub-folders within the current directory.

```
fs::dir_tree(path = "project_template", recurse = FALSE)
```

```
## project_template
## +-- data
## +-- data-raw
## +-- figs
## +-- output
## +-- readme.md
## \-- scripts
```

It's also possible to specify how many layers, by using an integer rather than `FALSE` or `TRUE`.

```
fs::dir_tree(path = "project_template", recurse = 2)
```

```
## project_template
## +-- data
## |   \-- readme.md
## +-- data-raw
## |   +-- readme.md
## |   \-- source_data.xlsx
## +-- figs
## |   \-- readme.md
## +-- output
## |   \-- readme.md
```

```
## +-- readme.md
## \-- scripts
##     \-- readme.md
```

Marwick, Boettinger, and Mullen provide some concrete examples of a file structure might look like; note that the structure explicitly follows that of an R package. Their figure showing a "medium compendium" is below (Marwick et al., 2018, p.84):

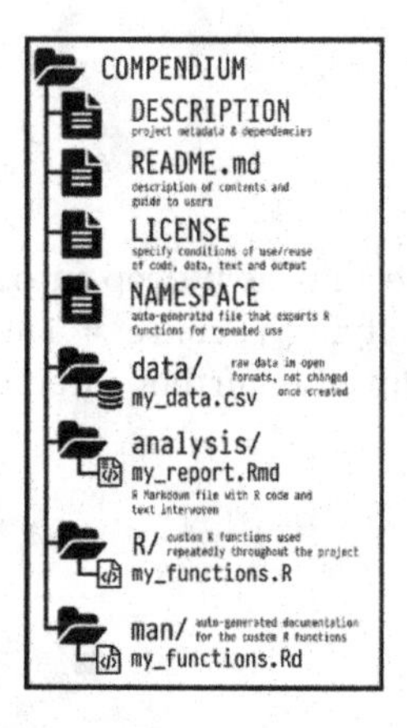

FIGURE 3.1 *Medium research compendium.*

In this example, there are elements that might not be relevant for every project (for example, the licenses for the code and data). But note that there's a clear separation of the data and the code that generates the analysis. As well, this structure is well-supported by the "project" approach, where what is shown here as the "COMPENDIUM" folder would be the root folder of your project. Everything we need is self-contained within this single folder.

3.4.3 The data dictionary or code book

As we will see, preparing data requires a number of steps, both objective and subjective, along the way. Creating a documentation of those decisions is essential. This documentation will serve as a record of the source of the data (all the better if you can also include specific information with the details of its collection such as dates, mode of collection, and the original survey questionnaire), the assumptions and decisions made in recoding and creating new variables, and so on.

Here are the three minimum components of a data dictionary:

1. Information about the variables (including units!) in the dataset not contained in the tidy data,
2. Information about the summary choices made, and
3. Information about the experimental study design [or the data source]

(Ellis and Leek, 2017)

Caitlin Hudon's "Field Notes: Building Data Dictionaries" (Hudon, 2018) provides some explicit guidelines as to what to include in a data dictionary, and poses "questions to answer with a data dictionary:"

1. What does the field mean? How should I use it?
2. What is the data supply chain?

- Where does this data come from? How exactly is it collected? How often is it updated? Where does it go next?

3. What does the data in this field actually look like?
4. Are there any caveats to keep in mind when using this data?
5. Where can I go to get more information?

Finally, the {labelled} package (Larmarange, 2022) contains the function `look_for()`, which generates a data dictionary of the dataframe that contains labelled variables"Generate a data dictionary and search for variables with look_for()"[4].

Here, we read the SPSS version of the {palmerpenguins} file, and then generate a data dictionary using the `look_for()` function. (We will return to using the {haven} package for reading data from SPSS, SAS, and Stata files later in this book.)

```
penguins_sav <- haven::read_sav(dpjr::dpjr_data("penguins.sav"))

labelled::look_for(penguins_sav)
```

```
##  pos variable          label col_type missing
##  1   species           –     dbl+lbl  0
##
##
```

[4]JosephLarmarange, http://larmarange.github.io/labelled/articles/look_for.html

```
## 2   island             -     dbl+lbl  0
##
##
## 3   bill_length_mm     -     dbl      2
## 4   bill_depth_mm      -     dbl      2
## 5   flipper_length_mm  -     dbl      2
## 6   body_mass_g        -     dbl      2
## 7   sex                -     dbl+lbl  11
##
## 8   year               -     dbl      0
## values
## [1] Adelie
## [2] Chinstrap
## [3] Gentoo
## [1] Biscoe
## [2] Dream
## [3] Torgersen
##
##
##
##
## [1] female
## [2] male
##
```

4

Importing data

In this chapter:

- Guidelines for importing data
- Setting the stage for subsequent chapters

4.1 Introduction

The data has been collected. It might be the result of:

- Running trials or experiments in a controlled laboratory setting.
- Making observations and recording the results of those observations (as an astronomer or biologist might).
- Sampling the people in an area and asking them a series of questions.
- Collecting information as part of a business operation. A familiar example are supermarket checkout scanners which collect data about grocery purchases. That data then facilitates operation management of the store by tracking sales and inventory and generating orders.

Once the data has been collected, one of the first steps in the analytic process is to import the data—read the contents of the file (or files) and begin preparing that data for the analysis. You might have collected the data yourself and designed the data storage. Or there may be a layer of processing between the raw source and the way it appears to you. After this processing, you might:

- Have access to a database where the raw data is stored, including where additional post-collection manipulation might happen,
- Be sent a file that contains a sample of the raw data,

- Be able to download data from a website, where the downloaded table has data that has already been compiled and summarized from a larger data set. An example would be a country's census data tables.

And it's important to note that your analysis project might require more than one of these methods of data collection.

How you assemble the data you need will depend on many factors, including what is already available, what your budget is (for example, some business-related data is collected by companies that then make it available at a cost), and the legal and regulatory environment (note that the definition of "personal information" varies from one jurisdiction to the next).

The specifics of the "import" step will differ by such things as the format of file and the variables, all of which we will explore in the next chapters.

4.2 Data formats

The data gets stored in a variety of electronic formats. The choice of format might be influenced by any one of the following:

- The underlying needs of the data collectors (some file formats are tailored to a specific use);
- The technology available to the collector;
- The nature of the data being collected.

There is sometimes (often?) no right answer as to the best format for a particular use case—there are pros and cons to each. (With that said, there is often a clearly *less good* choice for data storage and sharing—we're looking at you, PDF.) What this means is that in your workflow you will have to deal with data that needs to be extracted from a multitude of systems, and will be available to you in a multitude of formats.

> It is essential that a statistician can talk to the database specialist, and, as a team member, the statistician, along with most others, will be expected to be able to use the database facilities for most purposes by themselves, and of course advise on aspects of the design. There is always much preliminary 'data cleaning'

to do before an analysis can begin, almost regardless of how good a job is done by the database specialist. (Venables, 2010)

There are plenty of resources detailing the complexities of the different data storage formats, and the decision process that goes into determining which format is appropriate for a specific use-case. I always approach the task assuming that the professionals who built the data storage system made a well-informed decision, including balancing the various trade-offs between different formats, as well as budgetary and technology constraints that they might have faced.

4.3 Importing data

Here's some advice that's worth heeding:

1. The arguments in the import functions are your friends. Use them as your first line of defense in your project workflow.

"My main import advice: **use the arguments of your import function to get as far as you can, as fast as possible.** [*Emphasis added.*] Novice code often has a great deal of unnecessary post import fussing around. Read the docs for the import functions and take maximum advantage of the arguments to control the import." (Bryan and The STAT 545 TAs, 2019, Chapter 9: Writing and reading files)

In practice, this means that the first iteration of an import function will almost invariably not yield what you want. We will soon see some examples of this, but it might be in the variable names or how values are stored, how missing values are represented or the geometry of the spreadsheet. The arguments in the read function will allow you to address all of these issues, and more. Go back and add arguments to the function, and run the code again, and again, until you get to where you need to be.

2. Plain-text is boring but in the long run a more flexible and adaptable format; "Today's outputs are tomorrow's inputs" (Bryan and The STAT 545 TAs, 2019, Chapter 9: Writing and reading files)

"A plain text file that is readable by a human being in a text editor should be your default until you have **actual proof** that this will not work. Reading and writing to exotic or proprietary formats will be the first thing to break in the future or on a different computer. It also creates barriers for anyone who has a different toolkit than you do. Be software-agnostic. Aim for future-proof and moron-proof." (Bryan and The STAT 545 TAs, 2019, Chapter 9: Writing and reading files)

4.3.1 Check your results

It's always a good idea to quickly check your data after any major processing step. This can start with, but is not limited to, importing your data.

Some things to ask about the data:

- Did the import step give you as many records (rows) as you expected?
- Are there as many variables (columns) as you expected?
- Are the variable types (or classes) for those variables what you anticipated?
  - Did variables that should be numbers load as numeric types, or as character?
  - If working with labelled variables, did they load as factor type?
- How are missing values represented?
  - As we will see in the next chapters, many import functions provide the flexibility to encode NA values based on specific characters.

We will look at structured ways investigate the contents of the data in the chapter Validation strategies. In the next few chapters, we will focus on the process of reading the contents of different types of files.

4.3.2 Conclusion

In order to import data from different file types, we require different R functions, which we will delve into in the next chapters. While the specifics vary from one file type to the next, there are two general principles worth noting:

1. Use the arguments of your import function to get as far as you can, as fast as possible.
2. Check your results at the end of the import step—don't wait until later to discover that something went awry.

5

Importing data: plain-text files

In this chapter:

- Importing data from delimited text files
- Importing data from fixed-width text files

5.1 Delimited plain-text files

Plain-text files (sometimes called "ASCII files", after the character encoding standard they use) are often used to share data. They are limited in what they can contain, which has both downsides and upsides. On the downside, they can't carry any additional information with them, such as variable types and labels. But on the upside, they don't carry any additional information that requires additional interpretation by the software. This means they can be read consistently by a wide variety of software tools.

Plain-text files come in two varieties: delimited and fixed-width. "Delimited" is a reference to the fact that the files have a character that marks the boundary between two variables. A very common delimited format is the CSV file; the letters in the file name stand for "Comma Separated Values", and use a comma as the variable delimiter. Another delimited type, somewhat less common, uses the tab character to separate the variables, and will have the extension "TSV" for, you guessed it, "Tab Separated Values". Occasionally you will find files that use semi-colons, colons, or spaces as the delimiters.

There are base R functions to read this type of file. The functions within the {readr} (Wickham and Hester, 2020) package has some advantages over the base R functions when it comes to plain-text files. (The {readr} package is part of the tidyverse.) Compared to the equivalent base R functions, the {readr} functions are quite a bit faster, and the package's functions provide some useful flexibility when it comes to defining variable types as part of the read operation (rather than reading in the dat and then altering the variable types). As well, it returns a tibble instead of a dataframe. (For information about the difference, see (Wickham, 2019a, 3.6 Data frames and tibbles). For working with very large files, you may want to investigate the `fread()` function in the {data.table} package(Dowle et al., 2023).)

We activate {readr} by using the `library()` function:

```
library(readr)
```

In this chapter, we will be using two of the {readr} functions. Each has a variety of arguments that allow us to control the behaviour of the function; those will be dealt with in the examples.

function	purpose
`read_csv()`	reads the contents of a CSV file
`read_fwf()`	reads the contents of a fixed-width text file

5.2 Using {readr} to read a CSV file

In the following examples, we will use versions of the data in the {palmerpenguins} data (Horst et al., 2022; Horst, 2020)

In the code chunks below, an R character string object `penguins_path` is created that contains the text string with the location of the file "penguins.csv".

In the first example, the file has been loaded to our computer using the {dpjr} package, and uses the `dpjr_data()` function from that package to access the file (no matter its location on our computer).

In the second example, the file is located in a directory called "data" that is in our RStudio project folder. The {here} package will generate the full path, with slash or backslash separators (depending on your computer's operating system), starting at the location of the RStudio project.

The `penguins_path` object is then used in the function `read_csv()` to create a dataframe object called `penguins_data`.

```
# Example 2
# create object path using the {here} package
penguins_path <- here::here("data", "penguins.csv")
```

```
# read the contents of the CSV file
penguins_data <- read_csv(penguins_path)
```

```
Rows: 344 Columns: 8— Column specification
────────────────────────────────────────────────────────
Delimiter: ","
chr (3): species, island, sex
dbl (5): bill_length_mm, bill_depth_mm, flipper_length_mm, body_mass_g, year
ℹ Use `spec()` to retrieve the full column specification for this data.
ℹ Specify the column types or set `show_col_types = FALSE` to quiet this message.
```

FIGURE 5.1 *read_csv message.*

The function returns a message letting us know the type that each variable is assigned.

The arguments of the {readr} package allow a lot of control over how the file is read. Of particular utility are the following:

argument	purpose
`col_types = cols()`	define variable types
`na = ""`	specify which values you want to be turned into `NA`
`skip = 0`	specify how many rows to skip; the default is 0
`n_max = Inf`	the maximum number of records to read; the default is `Inf` for infinity, interpreted as the last row of the file

Adding the `col_types = cols()` parameter has two benefits. First, defining the variable type avoids errors (such as numeric fields will be read as such, and not as characters) and speeds up the read process since R doesn't have to check and infer the variable type.

Second, it allows us to alter what {readr} has decided for us. For example, we could set the `species` variable to be a factor type variable.

When we show the entire table, we can see that the variable `species` is now a factor type.

```
penguins_data <- read_csv(penguins_path,
                          col_types =
                            cols(species = col_factor()))
penguins_data
```

```
## # A tibble: 344 x 8
##    species island    bill_length_mm bill_depth_mm
##    <fct>   <chr>              <dbl>         <dbl>
##  1 Adelie  Torgersen           39.1          18.7
##  2 Adelie  Torgersen           39.5          17.4
##  3 Adelie  Torgersen           40.3          18
##  4 Adelie  Torgersen           NA            NA
##  5 Adelie  Torgersen           36.7          19.3
##  6 Adelie  Torgersen           39.3          20.6
##  7 Adelie  Torgersen           38.9          17.8
##  8 Adelie  Torgersen           39.2          19.6
##  9 Adelie  Torgersen           34.1          18.1
## 10 Adelie  Torgersen           42            20.2
## # i 334 more rows
## # i 4 more variables: flipper_length_mm <dbl>,
## #   body_mass_g <dbl>, sex <chr>, year <dbl>
```

If we were working with a very large file and wanted to read the first five rows, just to see what's there, we could use the `n_max =` argument:

```
read_csv(penguins_path,
         n_max = 5,
         show_col_types = FALSE   # option to turn off message
         )
```

```
## # A tibble: 5 x 8
##   species island    bill_length_mm bill_depth_mm
##   <chr>   <chr>              <dbl>         <dbl>
## 1 Adelie  Torgersen           39.1          18.7
## 2 Adelie  Torgersen           39.5          17.4
## 3 Adelie  Torgersen           40.3          18
## 4 Adelie  Torgersen           NA            NA
## 5 Adelie  Torgersen           36.7          19.3
## # i 4 more variables: flipper_length_mm <dbl>,
## #   body_mass_g <dbl>, sex <chr>, year <dbl>
```

5.3 Fixed-width files

Fixed-width files don't use a delimiter, and instead specify which column(s) each variable occupies, consistently for every row in the entire file.

Fixed-width files are a hold-over from the days when storage was expensive and/or on punch cards. This meant that specific columns in the table (or card) were assigned to a particular variable, and precious space was not consumed with a delimiter. Compression methods have since meant that a CSV file with unfixed variable lengths is more common, but in some big data applications, fixed-width files can be much more efficient.

If you ever have to deal with a fixed-width file, you will (or should!) receive a companion file letting you know the locations of each variable in every row.

In this example, we will use the one provided in the {dpjr} package, "authors_fwf.txt". This code chunk assigns the path to the file location, which we can use in our code later.

```
authors_path <- dpjr::dpjr_data("authors_fwf.txt")
```

This simple file has four (or as we will see, sometimes three, if we combine first and last name as one) variables, and three records (or rows).

- first name
- last name
- U.S. state of birth (two-letter abbreviation)
- randomly generated unique ID

If we open the file in a text editor, we see this:

```
Toni Morrison       IL          DJ-1944-QF96
Kurt Vonnegut       IN          XN-5632-TP58
Walt Whitman        NY          LW-1752-TD08
```

FIGURE 5.2 *authors_fwf.txt.*

We could also use the {base R} function `readLines()` to see the lines:

```
readLines(authors_path)
```

```
## [1] "Toni Morrison       IL          DJ-1944-QF96"
## [2] "Kurt Vonnegut       IN          XN-5632-TP58"
## [3] "Walt Whitman        NY          LW-1752-TD08"
```

The arguments within the `readr::read_fwf()` function include those listed above for the `read_csv()` function, and some others that are specific to fixed-width files.

argument	purpose
`col_names = c()`	defines a list of the names for the variables
`fwf_widths(widths = c(), col_names = c())`	a list of the character length of each variable, and their names
`fwf_positions(start = , end = , col_names = c())`	character position of the start and end of each variable, and their names
`fwf_cols(variable1_name = c(), variable2_name = c())`	name followed by start and end position of each variable)
`fwf_cols(variable1_name = width, variable2_name = width)`	name followed by width of each variable)

The examples below will elaborate on these arguments.

The first approach would be to allow {readr} to guess where the column breaks are. The `fwf_empty()` function looks through the specified file and returns the beginning and ending locations it has guessed, as well as the `skip` value that the `read_fwf()` function uses.

Note that the column names are specified in a list.

```
fwf_empty(
  authors_path,
  col_names = c("first", "last", "state", "unique_id")
  )
```

```
## $begin
## [1]  0  5 20 30
##
## $end
## [1]  4 13 22 NA
##
## $col_names
## [1] "first"     "last"      "state"     "unique_id"
```

That information can then be used by the `read_fwf()` function:

```
read_fwf(authors_path,
         fwf_empty(
           authors_path,
```

```
    col_names = c("first", "last", "state", "unique_id")
  ))
```

```
## # A tibble: 3 x 4
##   first last     state unique_id
##   <chr> <chr>    <chr> <chr>
## 1 Toni  Morrison IL    DJ-1944-QF96
## 2 Kurt  Vonnegut IN    XN-5632-TP58
## 3 Walt  Whitman  NY    LW-1752-TD08
```

Note that {readr} will impute the variable type, as it did with the CSV file. And although we won't implement it in these examples, in the same way `read_fwf()` allows us to use the `col_types` specification, as well as `na`, `skip`, and others. See the `read_fwf()` reference at https://readr.tidyverse.org/reference/read_fwf.html for all the details.

Reading this fixed-width file with these three author names worked, but it could break quite easily. We just need one person with three or more components to their name (initials, spaces, or hyphens, as in Ursula K. Le Guin[1] or Ta-Nehisi Coates[2]), or some missing values, and the inconsistent structure throws off the `read_fwf()` parser.

In the following example, we read a longer list of author names:

```
authors2_path <- dpjr::dpjr_data("authors2_fwf.txt")

fwf_empty(authors2_path,
          col_names = c("first", "last", "state", "unique_id"))
```

```
## $begin
## [1]  0 20 30
##
## $end
## [1] 19 22 NA
##
## $col_names
## [1] "first"     "last"      "state"     "unique_id"
```

The `fwf_empty()` function found only three columns, as shown in the "begin" and "end" values that are returned.

[1] https://en.wikipedia.org/wiki/Ursula_K._Le_Guin

[2] https://en.wikipedia.org/wiki/Ta-Nehisi_Coates

When we use the read function, it finds three columns:

```
read_fwf(dpjr::dpjr_data("authors2_fwf.txt"))
```

```
## # A tibble: 10 x 3
##    X1                  X2    X3
##    <chr>               <chr> <chr>
##  1 Toni Morrison       IL    DJ-1944-QF96
##  2 Kurt Vonnegut       IN    XN-5632-TP58
##  3 Walt Whitman        NY    LW-1752-TD08
##  4 Ursula K. Le Guin   CA    EZ-9789-EA77
##  5 Ta-Nehisi Coates    NY    YN-5151-NV82
##  6 W. E. B. Du Bois    MA    HN-6134-NF80
##  7 F. Scott Fitzgerald MN    YH-4405-TR02
##  8 N.K. Jemisin        IA    EF-7340-DW20
##  9 Flannery O'Connor   GA    HB-8269-XC88
## 10 Henry David Thoreau MA    RL-8200-SU83
```

```
Rows: 10 Columns: 3── Column specification
────────────────────────────────────────────────────────

chr (3): X1, X2, X3
i Use `spec()` to retrieve the full column specification for this data.
i Specify the column types or set `show_col_types = FALSE` to quiet this
message.
```

FIGURE 5.3 *read_fwf message.*

Adding the `col_names =` argument now mis-identifies the variables.

```
read_fwf(authors2_path,
         fwf_empty(
           authors2_path,
           col_names = c("first", "last", "state", "unique_id")
         ))
```

```
## # A tibble: 10 x 3
##    first               last  state
##    <chr>               <chr> <chr>
##  1 Toni Morrison       IL    DJ-1944-QF96
##  2 Kurt Vonnegut       IN    XN-5632-TP58
##  3 Walt Whitman        NY    LW-1752-TD08
##  4 Ursula K. Le Guin   CA    EZ-9789-EA77
```

```
## 5 Ta-Nehisi Coates     NY    YN-5151-NV82
## 6 W. E. B. Du Bois     MA    HN-6134-NF80
## 7 F. Scott Fitzgerald MN     YH-4405-TR02
## 8 N.K. Jemisin         IA    EF-7340-DW20
## 9 Flannery O'Connor    GA    HB-8269-XC88
## 10 Henry David Thoreau MA    RL-8200-SU83
```

```
Rows: 10 Columns: 3— Column specification
────────────────────────────────────────────────
chr (3): first, last, state
i Use `spec()` to retrieve the full column specification for this data.
i Specify the column types or set `show_col_types = FALSE` to quiet this
message.
```

FIGURE 5.4 *read_fwf message.*

A more reliable approach is to specify exactly the width of each column. Note that in the example below, we specify only "name" without splitting it into first and last.

The variables and their widths are as follows:

variable	width	start position	end position
name	20	1	20
state	10	21	30
uniqueID	12	31	42

The widths of each column can be added using the `fwf_widths` argument:

```
read_fwf(authors2_path,
         fwf_widths(widths = c(20, 10, 12),
                    col_names = c("name", "state", "unique_id")))
```

```
## # A tibble: 10 x 3
##   name                state unique_id
##   <chr>               <chr> <chr>
## 1 Toni Morrison       IL    DJ-1944-QF96
## 2 Kurt Vonnegut       IN    XN-5632-TP58
## 3 Walt Whitman        NY    LW-1752-TD08
## 4 Ursula K. Le Guin   CA    EZ-9789-EA77
## 5 Ta-Nehisi Coates    NY    YN-5151-NV82
## 6 W. E. B. Du Bois    MA    HN-6134-NF80
## 7 F. Scott Fitzgerald MN    YH-4405-TR02
## 8 N.K. Jemisin        IA    EF-7340-DW20
```

```
## 9 Flannery O'Connor   GA   HB-8269-XC88
## 10 Henry David Thoreau MA  RL-8200-SU83
```

```
Rows: 10 Columns: 3— Column specification

chr (3): X1, X2, X3
i Use `spec()` to retrieve the full column specification for this data.
i Specify the column types or set `show_col_types = FALSE` to quiet this
message.
```

FIGURE 5.5 *read_fwf message.*

A third option is to provide two lists of locations using `fwf_positions()`, the first with the start positions, and the second with the end positions. The first variable `name` starts at position 1 and ends at position 20, and the second variable `unique_id` starts at 30 and ends at 42. Note that we won't read the `state` variable which occupies the ten columns from 21 through 30.

```
read_fwf(authors2_path,
         fwf_positions(start = c(1, 31), end = c(20, 42),
                       col_names = c("name", "unique_id")))
```

```
## # A tibble: 10 x 2
##    name                unique_id
##    <chr>               <chr>
##  1 Toni Morrison       DJ-1944-QF96
##  2 Kurt Vonnegut       XN-5632-TP58
##  3 Walt Whitman        LW-1752-TD08
##  4 Ursula K. Le Guin   EZ-9789-EA77
##  5 Ta-Nehisi Coates    YN-5151-NV82
##  6 W. E. B. Du Bois    HN-6134-NF80
##  7 F. Scott Fitzgerald YH-4405-TR02
##  8 N.K. Jemisin        EF-7340-DW20
##  9 Flannery O'Connor   HB-8269-XC88
## 10 Henry David Thoreau RL-8200-SU83
```

```
Rows: 10 Columns: 2— Column specification

chr (2): name, unique_id
i Use `spec()` to retrieve the full column specification for this data.
i Specify the column types or set `show_col_types = FALSE` to quiet this
message.
```

FIGURE 5.6 *read_fwf message.*

The fourth option is a syntactic variation on the third, with the same values but in a different order. This time, all of the relevant information about each variable is aggregated, with the variable name followed by the start and end locations.

```
read_fwf(authors2_path,
         fwf_cols(name = c(1, 20), unique_id = c(31, 42)))
```

```
## # A tibble: 10 x 2
##    name                 unique_id
##    <chr>                <chr>
##  1 Toni Morrison        DJ-1944-QF96
##  2 Kurt Vonnegut        XN-5632-TP58
##  3 Walt Whitman         LW-1752-TD08
##  4 Ursula K. Le Guin    EZ-9789-EA77
##  5 Ta-Nehisi Coates     YN-5151-NV82
##  6 W. E. B. Du Bois     HN-6134-NF80
##  7 F. Scott Fitzgerald YH-4405-TR02
##  8 N.K. Jemisin         EF-7340-DW20
##  9 Flannery O'Connor    HB-8269-XC88
## 10 Henry David Thoreau RL-8200-SU83
```

```
Rows: 10 Columns: 2── Column specification
────────────────────────────────────────────────────────

chr (2): name, unique_id
i Use `spec()` to retrieve the full column specification for this data.
i Specify the column types or set `show_col_types = FALSE` to quiet this
message.
```

FIGURE 5.7 *read_fwf message.*

And finally, {readr} provides a fifth way to read in a fixed-width file that is a variation on the second approach we saw, with the name and the width values aggregated.

```
read_fwf(authors2_path,
         fwf_cols(name = 20, state = 10, unique_id = 12))
```

```
## # A tibble: 10 x 3
##    name                 state unique_id
##    <chr>                <chr> <chr>
##  1 Toni Morrison        IL    DJ-1944-QF96
##  2 Kurt Vonnegut        IN    XN-5632-TP58
```

```
##  3 Walt Whitman        NY    LW-1752-TD08
##  4 Ursula K. Le Guin   CA    EZ-9789-EA77
##  5 Ta-Nehisi Coates    NY    YN-5151-NV82
##  6 W. E. B. Du Bois    MA    HN-6134-NF80
##  7 F. Scott Fitzgerald MN    YH-4405-TR02
##  8 N.K. Jemisin        IA    EF-7340-DW20
##  9 Flannery O'Connor   GA    HB-8269-XC88
## 10 Henry David Thoreau MA    RL-8200-SU83
```

```
Rows: 10 Columns: 3— Column specification
────────────────────────────────────────────
chr (3): name, state, unique_id
i Use `spec()` to retrieve the full column specification for this data.
i Specify the column types or set `show_col_types = FALSE` to quiet this
message.
```

FIGURE 5.8 *read_fwf message.*

5.3.1 An extreme example of a fixed-width file

A particularly interesting research question is the relationship between education level and different health outcomes. In this example, we will start the process of importing a large file that contains data that will allow us to explore whether there is a correlation.

Statistics Canada has made available a Public-Use Microdata File (PUMF) of the Joint Canada/United States Survey of Health (JCUSH), a telephone survey conducted in late 2002 and early 2003. There were 8,688 respondents to the survey, 3,505 Canadians and 5,183 Americans. The data file that is available is anonymized, so we have access to the individual responses, which will facilitate additional analysis.

The webpage for the survey, including the PUMF file, data dictionary, and methodological notes, is here: https://www150.statcan.gc.ca/n1/pub/82m0022x/2003001/4069119-eng.htm

The PUMF is a fixed-width file named "JCUSH.txt". This file is quite a lot larger than the author names example above. There are the 8,688 records, one for each survey respondent. The data also consists of 366 variables, a combination of assigned variables (such as the unique respondent identification number and the country), survey question responses, and derived variables that are calculated as part of the post-survey data processing.

Here's what the first 40 characters of the first record looks like:

```
readLines(dpjr::dpjr_data("JCUSH.txt"), n = 1)
```

```
## [1] "1000033308351200211170432144421333431126"
```

There's not a bit of white space anywhere in this data file. The 366 variables are stored in only 552 columns—an average of 1.5 columns per variable! This is a good example of the efficiency associated with this approach to encoding the data.

Let's imagine that our research goal is to determine if there is a relationship between a person's level of education and their health outcomes. We've reviewed the data documentation, and it is clear that the JCUSH data has what we need.

Here's how one variable, highest level of post-secondary education achieved, appears in the data dictionary:

Variable Name	SDJ1GHED	**Length**	1	**Position**	502
Question Name					
Concept	Highest level of post-secondary education attained - (G)				
Question					
Universe	All respondents				
Note	Based on EDJ1_02.				

Content	**Code**	**Sample**	**Population**
LESS THAN HIGH SCHOOL	1	1,357	28,015,770
HIGH SCHOOL DEGREE OR EQUIVALENT (GED)	2	2,798	81,272,698
TRADES CERT,VOC. SCH./COMM.COL./CEGEP	3	1,451	32,707,506
UNIV OR COLL. CERT. INCL. BELOW BACH.	4	2,806	79,483,995
NOT STATED	9	276	8,984,053
	Total	8,688	230,464,023

FIGURE 5.9 *JCUSH health survey example.*

The variable `SDJ1GHED` is 1 character long, in position 502 of the data. The variable might be only 1 character long, but when coupled with the information in the Content table, it becomes a very powerful piece of information.

For our analysis question, we will read in four variables: the unique household identification number, the country, the overall health outcomes, and education level. For the variable names, we will use the same ones used in the data dictionary:

name	variable	length	position
SAMPLEID	Household identifier	12	1 - 12
SPJ1_TYP	Sample type [country]	1	13
GHJ1DHDI	Health Description Index	1	32
SDJ1GHED	Highest level of post-secondary education attained	1	502

You will note in the code below that in the case of the variables that are of length "1", the value to indicate the start and end positions are the same.

As well, the `SAMPLEID` variable is a 12-digit number; the `read_fwf()` function will interpret this as variable type double, and represent it in scientific notation. To be useful, we want to be able to see and evaluate the entire string. As a result, we use `col_types()` to specify `SAMPLEID` as a character.

```
jcush <- readr::read_fwf(dpjr::dpjr_data("JCUSH.txt"),
         fwf_cols(
           SAMPLEID = c(1, 12),
           SPJ1_TYP = c(13, 13),
           GHJ1DHDI = c(32, 32),
           SDJ1GHED = c(502, 502)
           ),
         col_types = list(
           SAMPLEID = col_character()
         ))

head(jcush)
```

```
## # A tibble: 6 x 4
##   SAMPLEID     SPJ1_TYP GHJ1DHDI SDJ1GHED
##   <chr>           <dbl>    <dbl>    <dbl>
## 1 100003330835        1        3        4
## 2 100004903392        1        3        2
## 3 100010137168        1        2        1
## 4 100010225523        1        3        3
## 5 100011623697        1        2        2
## 6 100013652729        1        4        3
```

Imagine, though, the challenge of handling this amount of data at one time! Between the many variables and the complex value labels, the "data" is more than just the fixed-width file. This is a circumstance where a different data storage solution (as we will see later) has some strengths.

6

Importing data: Excel

In this chapter:

- Importing data from Excel spreadsheets
- Extracting data that is encoded as Excel formatting

6.1 Introduction

Spreadsheets and similar data tables are perhaps the most common way that data is made available. Because they originated as an electronic version of accounting worksheets, they have a tabular structure that works for other situations when data collection, storage, analysis, and sharing (i.e. publishing) are required. Spreadsheet software of one form or another is often standard when you buy a new computer, and Microsoft Excel is the most common of all. Google also makes available a web-based spreadsheet tool, Google Sheets that works in much the same way.

While the data in a spreadsheet often gets stored in a manner similar to a plain-text file such as the CSV files we worked with in the previous chapter, Excel is designed to deliver far more functionality.

Broman and Woo (Broman and Woo, 2017) provide a how-to for good data storage practice in a spreadsheet, but you are much more likely to find yourself working with a spreadsheet that doesn't achieve that standard. Paul Murrell titled his article "Data Intended for Human Consumption, Not Machine Consumption" (Murrell, 2013) for a reason: all too often, a spreadsheet is used to make data easy for you and me to read, but this makes it harder for us to get it into a structure where our software program can do further analysis.

Spreadsheets like Excel also have a dark side (at least when it comes to data storage)—the values you see are not necessarily what's in the cell. For example, a cell might be the result of an arithmetic function that brings one or more values from elsewhere in the sheet (or in some cases, from another sheet, or anther sheet in another file). Some users will colour-code cells, but with no index to tell you what each colour means. (Bryan, 2016b) (For an alternative vision, see "Sanesheets" (Bryan, 2016a).)

The "format as data" and "formatting as information" approaches of storing information can make it hard for us to undertake our analysis. Excel files can also contain a wide variety of data format types. For example, when using Excel you will find that leading or trailing whitespace makes a number turn into a character, even if the cell is given an Excel number format.

The {readxl} package (Wickham and Bryan, 2019) is designed to solve many of the challenges of reading Excel files. To access the functions in {readxl}, we use the `library()` function to load the package into our active project:

```
library(readxl)
```

{readxl} tries to figure out what's going on with each variable, but like {readr} it allows you to override some of those automated decisions if it returns something different than what the analysis calls for. The {readxl} arguments we will use include the following:

argument	purpose
`sheet = ""`	a character string with the name of the sheet in the Excel workbook to be read
`range = ""`	define the rectangle to be read; see the examples below for syntax options
`skip = 0`	specify how many rows to skip; the default is 0
`n_max = Inf`	the maximum number of records to read; the default is `Inf` for infinity, interpreted as the last row of the file

For this example, we will use some mock data that is the type a data analyst might encounter when undertaking analysis of human resource data to support business decision-making. The file "cr25_human_resources.xlsx" is in the folder "cr25".

Because Excel files can contain multiple sheets, we need to specify which one we want to read. The {readxl} function `excel_sheets()` lists them for us:

```
excel_sheets(dpjr::dpjr_data("cr25/cr25_human_resources.xlsx"))
```

```
## [1] "readme"            "df_HR_main"
## [3] "df_HR_transaction" "df_HR_train"
```

The one we want for our example is "df_HR_main", which contains the employee ID, name, birth date, and gender for each of the 999 staff.

In our `read_excel()` function we can specify either the name of the sheet or the number based on its position in the Excel file. While the default is to read the first sheet, it's often good practice to be specific, as Excel users are prone to re-arranging the order of the sheets (including adding new sheets). And because they seem (slightly) less likely to rename sheets, we have another reason to use the name rather than the number.

```
df_HR_main <-
  read_excel(
    path = dpjr::dpjr_data("cr25/cr25_human_resources.xlsx"),
    sheet = "df_HR_main"
    )
```

```
## New names:
## * `` -> `...2`
## * `` -> `...3`
## * `` -> `...4`
## * `` -> `...5`
## * `` -> `...6`
## * `` -> `...7`
```

```
head(df_HR_main)
```

```
## # A tibble: 6 x 7
##   df_HR_main          ...2  ...3  ...4  ...5  ...6  ...7
##   <chr>               <chr> <chr> <chr> <chr> <lgl> <chr>
## 1 The main databas~ <NA>  <NA>  <NA>  <NA>  NA    <NA>
## 2 <NA>                <NA>  <NA>  <NA>  <NA>  NA    <NA>
## 3 <NA>                <NA>  <NA>  <NA>  <NA>  NA    <NA>
## 4 <NA>                emp_~ date~ name  gend~ NA    NOTE:
## 5 <NA>                ID001 29733 Avri~ Fema~ NA    date~
## 6 <NA>                ID002 26931 Katy~ Fema~ NA    <NA>
```

This sheet possesses a structure we commonly see in Excel files: in addition to the data of interest, the creators of the file have helpfully added a title for the sheet. They might also put a note in the "margins" of the data (often to the right or below the data), and add formatting to make it more visually appealing to human readers (beware empty or hidden columns and rows!).

While this is often useful information, its presence means that we need to trim rows and columns to get our dataframe in a useful structure for our analysis.

The `read_excel()` function provides a number of different ways for us to define the range that will be read. One that works well in the function, and aligns with how many Excel users think about their spreadsheets, defines the upper left and the lower right corners of the data using Excel's column-row alpanumeric nomenclature. For this file, that will exclude the top rows, the empty columns to the left and right of the data, as well as the additional note to the right. The corners are cells B5 and E1004. How did I know that?

Important note: there is no shame in opening the file in Excel to make a visual scan of the file, looking for anomalous values and to note the corners of the data rectangle.

```
df_HR_main <-
  read_excel(
    path = dpjr::dpjr_data("cr25/cr25_human_resources.xlsx"),
    sheet = "df_HR_main",
    range = "B5:E1004"
    )

head(df_HR_main)
```

```
## # A tibble: 6 x 4
##   emp_id date_of_birth       name              gender
##   <chr>  <dttm>              <chr>             <chr>
## 1 ID001  1981-05-27 00:00:00 Avril Lavigne     Female
## 2 ID002  1973-09-24 00:00:00 Katy Perry        Female
## 3 ID003  1966-11-29 00:00:00 Chester Bennington Male
## 4 ID004  1982-09-17 00:00:00 Chris Cornell     Male
## 5 ID005  1976-07-17 00:00:00 Bryan Adams       Male
## 6 ID006  1965-07-03 00:00:00 Courtney Love     Female
```

The `range =` argument also gives us the option to use Excel's row number-column number specification, which would change the above `range = "B5:E1004"` to `range = "R5C2:R1004C5"` ("B" refers to the second column, and "E" to the fifth).

```
df_HR_main <-
  read_excel(
```

```
    path = dpjr::dpjr_data("cr25/cr25_human_resources.xlsx"),
    sheet = "df_HR_main",
    range = "R5C2:R1004C5"
    )
```

There are a variety of other approaches to defining the data rectangle in a `read_excel()` function, which might be more appropriate in a different context. For example, if the sheet is continually having new rows added at the bottom, using the `range = anchored("B5", dim =c(NA, 4))` specification might be prefered. This will start at cell B5, then read as many rows as there are data (the "NA" inside the `c()` function and four columns[1].

```
df_HR_main <-
  read_excel(
    path = dpjr::dpjr_data("cr25/cr25_human_resources.xlsx"),
    sheet = "df_HR_main",
    range = anchored("B5", dim =c(NA, 4))
    )

head(df_HR_main)
```

```
## # A tibble: 6 x 4
##   emp_id date_of_birth       name              gender
##   <chr>  <dttm>              <chr>             <chr>
## 1 ID001  1981-05-27 00:00:00 Avril Lavigne     Female
## 2 ID002  1973-09-24 00:00:00 Katy Perry        Female
## 3 ID003  1966-11-29 00:00:00 Chester Bennington Male
## 4 ID004  1982-09-17 00:00:00 Chris Cornell     Male
## 5 ID005  1976-07-17 00:00:00 Bryan Adams       Male
## 6 ID006  1965-07-03 00:00:00 Courtney Love     Female
```

Another splendid feature of `read_excel()` is that it returns a date type (S3: POSIXct) for variables that it interprets as dates.

[1] For more options, see https://readxl.tidyverse.org/articles/sheet-geometry.html

6.2 Extended example

Very often, Excel files are far more complicated than the case above. In the following example, we will read the contents of one sheet in an Excel file published by the UK Office of National Statistics, drawn from the 2021 Census of England and Wales. The file "census2021firstresultsenglandwales1.xlsx" has the population by local authorities (regions), and includes sheets with notes about the data, and data tables containing the population by sex and five-year age groups, along with population density and number of households.

Our focus is on sheet "P01", with population by sex. The first few rows of the data look like this:

	A	B	C	D	E	F	G	H
1	**P01 Census 2021: Usual resident population by sex, local authorities in England and Wales [note 1]**							
2	**England and Wales: regions (within England), unitary authorities, counties, districts, London boroughs**							
3	This worksheet contains one dataset.							
4	Notes for the data can be found in the notes worksheet.							
5	Source: Office for National Statistics							
6	Released: 28 June 2022							
7	**Area code [note 2]**	**Area name**	**All persons**	**Females**	**Males**			
8	K04000001	**England and Wales**	59,597,300	30,420,100	29,177,200			
9	E92000001	**England**	56,489,800	28,833,500	27,656,300			
10	E12000001	**North East**	2,647,100	1,353,800	1,293,300			
11	E06000047	County Durham	522,100	266,800	255,300			
12	E06000005	Darlington	107,800	55,100	52,700			
13	E06000001	Hartlepool	92,300	47,700	44,700			
14	E06000002	Middlesbrough	143,900	73,000	70,900			
15	E06000057	Northumberland	320,600	164,000	156,600			
16	E06000003	Redcar and Cleveland	136,500	70,500	66,100			
17	E06000004	Stockton-on-Tees	196,600	100,100	96,500			
18	E11000007	Tyne and Wear (Met County)	1,127,200	576,600	550,700			
19	E08000037	Gateshead	196,100	100,000	96,200			
20	E08000021	Newcastle upon Tyne	300,200	151,800	148,400			
21	E08000022	North Tyneside	209,000	107,600	101,400			
22	E08000023	South Tyneside	147,800	76,100	71,700			
23	E08000024	Sunderland	274,200	141,100	133,100			
24	E12000002	**North West**	7,417,300	3,777,200	3,640,100			
25	E06000008	Blackburn with Darwen	154,800	78,100	76,800			
26	E06000009	Blackpool	141,100	71,400	69,700			
27	E06000049	Cheshire East	398,800	203,200	195,500			
28	E06000050	Cheshire West and Chester	357,200	182,800	174,400			
29	E06000006	Halton	128,200	65,500	62,800			
30	E06000007	Warrington	210,900	106,500	104,400			
31	E10000006	Cumbria	499,800	253,700	246,100			
32	E07000026	Allerdale	96,100	49,100	47,100			
33	E07000027	Barrow-in-Furness	67,400	34,000	33,400			

FIGURE 6.1 *UK Census data.*

The {readxl} package's `read_excel()` function is the one we want to use. The first item is the path to the file. The function's default is to read the first sheet in the Excel file, which is not what we want, so we use the `sheet =` argument to specify "P01". We will also read only the first 10 rows in the sheet, using the `n_max = 10` specification.

```
uk_census_pop <- readxl::read_excel(
  dpjr::dpjr_data("census2021firstresultsenglandwales1.xlsx"),
  sheet = "P01",
  n_max = 10
)
```

```
uk_census_pop
```

A tibble: 10 x 5

P01 Census 2021: Usual resident population by sex, local authorities in England and Wales [note 1]
<chr>

England and Wales: regions (within England), unitary authorities, counties, districts, London boroughs
This worksheet contains one dataset.
Notes for the data can be found in the notes worksheet.
Source: Office for National Statistics
Released: 28 June 2022
Area code [note 2]
K04000001
E92000001
E12000001
E06000047

1-10 of 10 rows | 1-1 of 5 columns

FIGURE 6.2 *UK Census data.*

Our visual inspection had already informed us that the first few rows at the top of the spreadsheet contain the title (now in the variable name row), along with things like the source and the release date. While this documentation is important, it's not part of the data we want for our analysis.

Another of the `read_excel()` arguments to specify the rectangular geometry that contains our data is `skip =`. Because our visual inspection showed us that the header row for the data of interest starts at the seventh row, we will use it to skip the first 6 rows.

```
uk_census_pop <- readxl::read_excel(
  dpjr::dpjr_data("census2021firstresultsenglandwales1.xlsx"),
  sheet = "P01",
  n_max = 10,
  skip = 6
)

head(uk_census_pop)
```

```
## # A tibble: 6 x 5
##   `Area code [note 2]` `Area name`        `All persons`
##   <chr>                <chr>                      <dbl>
## 1 K04000001            England and Wales       59597300
## 2 E92000001            England                 56489800
## 3 E12000001            North East               2647100
## 4 E06000047            County Durham             522100
## 5 E06000005            Darlington                107800
## 6 E06000001            Hartlepool                 92300
## # i 2 more variables: Females <dbl>, Males <dbl>
```

A second option would be to define the range of the cells, using Excel nomenclature to define the upper-left and bottom-right of the data rectangle. This is often a good approach when there is content outside the boundary of the data rectangle, such as additional notes.

Again, opening the file in Excel for a visual scan can provide some valuable insights...here, we can see the additional rows at the top, as well as the hierarchical nature of the data structure.

In this code, we will also apply the `clean_names()` function from the {janitor} package (Firke, 2021), which cleans the variable names. Specifically, it removes the spaces (and any other problematic characters), and puts the variable names into lower case. Finally, the `rename()` function is applied to remove the "_note_2" text from the area code variable.

```
uk_census_pop <- readxl::read_excel(
  dpjr::dpjr_data("census2021firstresultsenglandwales1.xlsx"),
  sheet = "P01",
  range = "A7:E382"
) |>
  janitor::clean_names() |>
  rename(area_code = area_code_note_2)

uk_census_pop
```

```
## # A tibble: 375 x 5
##    area_code area_name         all_persons females   males
##    <chr>     <chr>                   <dbl>   <dbl>   <dbl>
##  1 K04000001 England and Wa~      59597300  3.04e7 2.92e7
##  2 E92000001 England              56489800  2.88e7 2.77e7
##  3 E12000001 North East            2647100  1.35e6 1.29e6
##  4 E06000047 County Durham          522100  2.67e5 2.55e5
##  5 E06000005 Darlington             107800  5.51e4 5.27e4
##  6 E06000001 Hartlepool              92300  4.77e4 4.47e4
```

```
## 7 E06000002 Middlesbrough       143900  7.3 e4 7.09e4
## 8 E06000057 Northumberland      320600  1.64e5 1.57e5
## 9 E06000003 Redcar and Cle~     136500  7.05e4 6.61e4
## 10 E06000004 Stockton-on-Te~    196600  1.00e5 9.65e4
## # i 365 more rows
```

The data has been read correctly, including the fact that the data variables are read in the correct format. But this loses an important piece of information: the geographical hierarchy. In the Excel file, the levels of the hierarchy are represented by the indentation. Our next challenge is to capture that information as data we can use.

6.2.1 Excel formatting as data

One thing to notice about this data is the `Area name` variable is hierarchical, with a series of sub-totals. The lower (smaller) levels, containing the component areas of the next higher level in the hierarchy, are represented by indentations in the area name value. The lower the level in the hierarchy, the more indented the name of the area.

The specifics of the hierarchy are as follows:

- The value "England and Wales" contains the total population, which is the sum of England (Excel's row 8) and Wales (row 360).
- England (but not Wales) is then further subdivided into nine sub-national Regions (coded as "E12" in the first three characters of the `Area code` variable).
- These Regions are then are divided into smaller units, but the description of those units varies. These are Unitary Authorities (E06), London Boroughs (E09), Counties (E10), and Metropolitan Counties (E11); for our purpose, let's shorthand this level as "counties".
- The county level is further divided into Non-Metropolitan Districts (E07), and Metropolitan Counties are subdivided into Metropolitan Districts (E08). The image above shows Tyne and Wear (an E11 Met County, at row 18) is subdivided, starting with Gateshead, while Cumbria (an E10 County, at row 31) is also subdivided, starting with Allerdale.
- Wales is divided into Unitary Authorities, coded as "W06" in the first three characters of the `Area code` variable. In Wales, there is no intervening Region, so these are at the same level as the Unitary Authorities in England[2]. "Local

[2] The hierarchy of the administrative geographies of the UK can be found at the "England" and "Wales" links on [the ONS "Administrative geographies" page] (https://www.ons.gov.uk/methodology/geography/ukgeographies/administrativegeography).

Authority Districts, Counties and Unitary Authorities (April 2021) Map in United Kingdom"[3].

This complicated structure means that the data has multiple subtotals—a `sum()` function on the `All persons` variable will lead to a total population of 277,092,700, roughly 4.7 times the actual population of England and Wales in 2021, which was 59,597,300.

What would be much more useful is a structure that has the following characteristics:

- No duplication (people aren't double-counted).
- A code for the level of the hierarchy, to enable filtering.
- Additional variables (columns), so that each row has the higher level area of which it is a part. For example, County Durham would have a column that contains the regional value "North East" and the country value of "England".

With that structure, we could use the different regional groupings in a `group_by()` function.

As we saw earlier, the {readxl} function `read_excel()` ignores the formatting, so there is no way to discern the level in the characters.

Let's explore approaches to extracting the information that is embedded in the data and the formatting, and use that to prune the duplicated rows.

Approach 1 - use the Area code

The first approach we can take is to extract the regional element in the `area_code` variable, and use that to filter the data.

For example, if we want to plot the population in England by the county level, an approach would be to identify all those rows that have an Area code that starts with one of the following:

- Unitary Authorities (E06), London Boroughs (E09), Counties (E10), and Metropolitan Counties (E11)

For this, we can design a regular expression that identifies a letter "E" at the beginning by using the "^" character. (Remember, the "E" is for England, and "W" for Wales.) This is then followed by either a 0 or a 1 in the 2nd character

[3]You can download a map of these regions at https://geoportal.statistics.gov.uk/documents/ons::local-authority-districts-counties-and-unitary-authorities-april-2021-map-in-united-kingdom–1/about

spot (within the first set of square brackets), and in the third spot one of 6, 9, 0, or 1 (within the second set of square brackets).

```
county_pop <- uk_census_pop |>
  filter(str_detect(area_code, "^E[01][6901]"))

dplyr::glimpse(county_pop, width = 65)
```

```
## Rows: 122
## Columns: 5
## $ area_code   <chr> "E06000047", "E06000005", "E06000001", "E06~
## $ area_name   <chr> "County Durham", "Darlington", "Hartlepool"~
## $ all_persons <dbl> 522100, 107800, 92300, 143900, 320600, 1365~
## $ females     <dbl> 266800, 55100, 47700, 73000, 164000, 70500,~
## $ males       <dbl> 255300, 52700, 44700, 70900, 156600, 66100,~
```

We can check to ensure that the regex has worked as intended through the `distinct()` function, examining only the first three characters of the `area_code` strings:

```
county_pop  |>
  distinct(str_sub(area_code, 1, 3))
```

```
## # A tibble: 4 x 1
##   `str_sub(area_code, 1, 3)`
##   <chr>
## 1 E06
## 2 E11
## 3 E10
## 4 E09
```

Another approach would be to create a list of the desired three-character codes, then create a new variable in the main data, and filter against a list of UA codes:

```
# create list of UA three-character codes
county_list <- c("E06", "E09", "E10", "E11")

county_pop <- uk_census_pop |>
  # create three character area prefix
  mutate(area_code_3 = str_sub(area_code, 1, 3)) |>
  # filter that by the county_list
  filter(area_code_3 %in% county_list)
```

For our validation checks, we will ensure that our list of UAs is complete, and that the calculated total population matches what is given in the "England" row of the source data:

```
# check
county_pop  |>
  distinct(area_code_3)
```

```
## # A tibble: 4 x 1
##   area_code_3
##   <chr>
## 1 E06
## 2 E11
## 3 E10
## 4 E09
```

```
# check
county_pop  |>
  summarize(total_pop = sum(all_persons))
```

```
## # A tibble: 1 x 1
##   total_pop
##       <dbl>
## 1  56489600
```

While this approach works for our particular example, it will require careful coding every time we want to use it. Because of that limited flexibility, we may want to use another approach.

Approach 2 - use Excel's formatting

The leading space we see when we view the Excel file in its native software is not created by space characters, but through Excel's "indent" formatting. The {tidyxl} package (Garmonsway et al., 2022) has functions that allow us to turn that formatting into information, which we can then use in the functions within the {unpivotr} package (Garmonsway, 2023).

```
library(tidyxl)
library(unpivotr)
```

The code below uses the function `xlsx_sheet_names()` to get the names of all of the sheets in the Excel file.

```
tidyxl::xlsx_sheet_names(
  dpjr::dpjr_data("census2021firstresultsenglandwales1.xlsx")
  )
```

```
## [1] "Cover sheet" "Contents"     "Notes"
## [4] "P01"         "P02"          "P03"
## [7] "P04"         "H01"
```

We already know that Excel files can contain a lot of information that is outside the data rectangle. Rather than ignore all the information that's in the sheet, we will use the function `tidyxl::xlsx_cells()` to read the entire sheet. This uses the object with the file path character string and the `sheets =` argument to specify which one to read. Unlike `readxl::read_excel()`, we won't specify the sheet geometry. This will read the contents of the entire sheet, and we will extract the data we want from the object `uk_census` created at this step.

```
uk_census <- tidyxl::xlsx_cells(
  dpjr::dpjr_data("census2021firstresultsenglandwales1.xlsx"),
  sheets = "P01"
  )

# take a quick look
dplyr::glimpse(uk_census, width = 65)
```

```
## Rows: 1,888
## Columns: 24
```

```
## $ sheet               <chr> "P01", "P01", "P01", "P01", "P01", ~
## $ address             <chr> "A1", "D1", "A2", "A3", "A4", "A5",~
## $ row                 <int> 1, 1, 2, 3, 4, 5, 6, 6, 7, 7, 7, 7,~
## $ col                 <int> 1, 4, 1, 1, 1, 1, 1, 4, 1, 2, 3, 4,~
## $ is_blank            <lgl> FALSE, TRUE, FALSE, FALSE, FALSE, F~
## $ content             <chr> "97", NA, "98", "99", "100", "101",~
## $ data_type           <chr> "character", "blank", "character", ~
## $ error               <chr> NA, NA, NA, NA, NA, NA, NA, NA, NA,~
## $ logical             <lgl> NA, NA, NA, NA, NA, NA, NA, NA, NA,~
## $ numeric             <dbl> NA, NA, NA, NA, NA, NA, NA, NA, NA,~
## $ date                <dttm> NA, NA, NA, NA, NA, NA, NA, NA, NA~
## $ character           <chr> "P01 Census 2021: Usual resident po~
## $ character_formatted <list> [<tbl_df[1 x 14]>], <NULL>, [<tbl_~
## $ formula             <chr> NA, NA, NA, NA, NA, NA, NA, NA, NA,~
## $ is_array            <lgl> FALSE, FALSE, FALSE, FALSE, FALSE, ~
## $ formula_ref         <chr> NA, NA, NA, NA, NA, NA, NA, NA, NA,~
## $ formula_group       <int> NA, NA, NA, NA, NA, NA, NA, NA, NA,~
## $ comment             <chr> NA, NA, NA, NA, NA, NA, NA, NA, NA,~
## $ height              <dbl> 19.15, 19.15, 16.90, 15.00, 15.00, ~
## $ width               <dbl> 21.71, 12.43, 21.71, 21.71, 21.71, ~
## $ row_outline_level   <dbl> 1, 1, 1, 1, 1, 1, 1, 1, 1, 1, 1, 1,~
## $ col_outline_level   <dbl> 1, 1, 1, 1, 1, 1, 1, 1, 1, 1, 1, 1,~
## $ style_format        <chr> "Heading 1", "Normal", "Heading 2",~
## $ local_format_id     <int> 48, 29, 30, 14, 14, 14, 14, 14, 55,~
```

The dataframe created by the `xlsx_cells()` function is nothing like the spreadsheet in the Excel file. Instead, every cell in the Excel file is a row, and details about that cell are captured. The cell location is captured in the variable "address", while there is also a separate variable for the row number and another for the column number. There is also the value we would see if we were to look at the file, in the variable `character`. (If there is a formula in a cell, the function returns the result of the formula in the `character` variable, and the text of the formula in the `formula` variable.)

The `local_format_id` variable is created by {tidyxlr}, and helps us solve the problem of capturing the indentation. This variable contains a unique value for every different type of formatting that is used in the sheet. Below, we look at the first ten rows of the `uk_census` object, and we can see that the value "14" appears repeatedly in the variable "local_format_id".

```
uk_census |>
  select(address, character, local_format_id) |>
  glimpse(width = 65)
```

```
## Rows: 1,888
## Columns: 3
## $ address         <chr> "A1", "D1", "A2", "A3", "A4", "A5", "A6~
## $ character       <chr> "P01 Census 2021: Usual resident popula~
## $ local_format_id <int> 48, 29, 30, 14, 14, 14, 14, 14, 55, 55,~
```

The next thing we need to do is read the details of each of the formats stored in the Excel file—the format "14" has particular characteristics, including the number of indentations in the cell.

For this, we use the {tidyxl} function `xlsx_formats()`. We only need to specify the path to the file. "Excel defines only one of these dictionaries for the whole workbook, so that it records each combination of formats only once. Two cells that have the same formatting, but are in different sheets, will have the same format code, referring to the same entries in that dictionary"[4].

The code below reads the formats and assigns the results to a new object `uk_census_formats`.

```
uk_census_formats <- xlsx_formats(
  dpjr::dpjr_data("census2021firstresultsenglandwales1.xlsx"),
)
```

When examining the `uk_census_formats` object in our environment, we see that it contains two lists, one called "local" and the other "style". If we look inside "local", we see one called "alignment", and within that is "indent"—this is what we're looking for.

You may have noticed that there are 62 elements in these lists, not 382 (the number of rows in the Excel file), or the 1,888 that represents each active cell in the sheet P01. What the 62 represents is the number of different formats present in the entire Excel file (across all of the sheets).

We can extract that list-within-a-list, which contains all the indentation specifications of the different cell formats in the file, using the code below:

```
indent <- uk_census_formats$local$alignment$indent
indent
```

```
##  [1] 0 0 0 0 0 0 0 0 0 0 0 0 0 0 0 0 0 0 0 0 0 0 0 0 0
## [26] 0 0 0 0 0 0 0 0 0 0 0 0 0 0 0 0 0 0 0 0 0 0 0 0 0
## [51] 0 0 0 0 0 0 0 1 2 4 3 0
```

[4]Duncan Garmonsway, response to issue #84, {tidyxl} package repository, 2022-05-26. https://github.com/nacnudus/tidyxl/issues/84

Name	Type	Value
uk_census_formats	list [2]	List of length 2
local	list [6]	List of length 6
numFmt	character [62]	'General' 'General' 'General' 'General' 'General' 'General' ...
font	list [10]	List of length 10
fill	list [2]	List of length 2
border	list [12]	List of length 12
alignment	list [8]	List of length 8
horizontal	character [62]	'general' 'general' 'general' 'general' 'general' 'general' ...
vertical	character [62]	'bottom' 'bottom' 'bottom' 'bottom' 'bottom' 'bottom' ...
wrapText	logical [62]	FALSE TRUE TRUE TRUE TRUE TRUE ...
readingOrder	character [62]	'context' 'context' 'context' 'context' 'context' 'context' ...
indent	integer [62]	0 0 0 0 0 0 ...
justifyLastLine	logical [62]	FALSE FALSE FALSE FALSE FALSE FALSE ...
shrinkToFit	logical [62]	FALSE FALSE FALSE FALSE FALSE FALSE ...
textRotation	integer [62]	0 0 0 0 0 0 ...
protection	list [2]	List of length 2
style	list [6]	List of length 6

FIGURE 6.3 *UK Census data.*

This shows that for most of the various cell formattings used in the file most have zero indentation, but some have 1, 2, 3, or 4 indents.

Back to the object `uk_census`. We are interested in the data starting in the eighth row, so we need to filter for that.

For the {tidyxl} functions to work, the first column needs to define the hierarchical structure. To accomplish this, we will select it out using the `!=` operator.

The `behead_if()` function identifies a level of headers in a pivot table, and makes it part of the data. Similar to the {tidyr} function `pivot_longer()`, it creates a new variable for each row based on the headers. In the case of the UK Census data, the headers are differentiated by the number of indents in the formatting.

Let's test that by assigning the "left-up" position value to a new variable, field0. For each row in the `uk_census` object, the number of indentations associated with the `local_format_id` is evaluated; if it's zero, the following happens: starting at the numeric value, the function looks left and then up until it finds a value that has fewer indentations. For example, the "England" row has a single indentation, so the function looks left and up until it finds a value that has zero indents in the formatting; the value from that cell is assigned to the variable `field0` (for zero indents).

```
uk_census |>
  dplyr::filter((row >= 7)) |>
  filter(col != 1) |>
  behead_if(indent[local_format_id] == 0,
            direction = "left-up",
            name = "field0"
            ) |>
  select(address, row, col, content, field0) |>
  dplyr::filter(row < 30)
```

```
## # A tibble: 90 x 5
##    address   row   col content  field0
##    <chr>   <int> <int> <chr>    <chr>
##  1 C7          7     3 104      Area name
##  2 D7          7     4 105      Area name
##  3 E7          7     5 106      Area name
##  4 C8          8     3 59597300 England and Wales
##  5 D8          8     4 30420100 England and Wales
##  6 E8          8     5 29177200 England and Wales
##  7 B9          9     2 110      England and Wales
##  8 C9          9     3 56489800 England and Wales
##  9 D9          9     4 28833500 England and Wales
## 10 E9          9     5 27656300 England and Wales
## # i 80 more rows
```

Let's continue that through all the levels of indentation. There are four, so we will create a variable that starts with `field` and then a digit with the number of indentations.

```
uk_census_behead <- uk_census |>
  dplyr::filter((row >= 7)) |>
  filter(col != 1) |>
  behead_if(indent[local_format_id] == 0,
            direction = "left-up", name = "field0") |>
  behead_if(indent[local_format_id] == 1,
            direction = "left-up", name = "field1") |>
  behead_if(indent[local_format_id] == 2,
            direction = "left-up", name = "field2") |>
  behead_if(indent[local_format_id] == 3,
            direction = "left-up", name = "field3") |>
  behead_if(indent[local_format_id] == 4,
            direction = "left-up", name = "field4")
```

In the version below, only the final line changes. All of the other indentation values have been evaluated, so it doesn't require the `_if` and the associated evaluation, nor does it require the "-up" of the direction.

```
uk_census_behead <- uk_census |>
  dplyr::filter((row >= 7)) |>
  filter(col != 1) |>
  behead_if(indent[local_format_id] == 0,
            direction = "left-up", name = "field0") |>
  behead_if(indent[local_format_id] == 1,
            direction = "left-up", name = "field1") |>
  behead_if(indent[local_format_id] == 2,
            direction = "left-up", name = "field2") |>
  behead_if(indent[local_format_id] == 3,
            direction = "left-up", name = "field3") |>
  behead(direction = "left", name = "field4")
```

In the code below, the addition is the assignment of column headers. In the UK Census data, the structure is not particularly complex, but {tidyr} can also deal with nested hierarchy in the headers as well.

```
# from the previous chunk:

uk_census_behead <- uk_census |>
  dplyr::filter((row >= 7)) |>
  filter(col != 1) |>
  behead_if(indent[local_format_id] == 0,
            direction = "left-up", name = "field0") |>
  behead_if(indent[local_format_id] == 1,
            direction = "left-up", name = "field1") |>
  behead_if(indent[local_format_id] == 2,
            direction = "left-up", name = "field2") |>
  behead_if(indent[local_format_id] == 3,
            direction = "left-up", name = "field3") |>
  behead(direction = "left", name = "field4") |>
#  now strip (behead) the column names
  behead(direction = "up", name = "gender") |>
# add row sorting to preserve visual comparability
  arrange(row)
```

Did we get the results we expected? Let's do some quick checks of the data, first to see if the 1,125 cells of Excel data (CX to DY) are represented by an individual row (the result of the pivot):

```
glimpse(uk_census_behead, width = 65)
```

```
## Rows: 1,125
## Columns: 30
## $ sheet               <chr> "P01", "P01", "P01", "P01", "P01", ~
## $ address             <chr> "C8", "D8", "E8", "C9", "D9", "E9",~
## $ row                 <int> 8, 8, 8, 9, 9, 9, 10, 10, 10, 11, 1~
## $ col                 <int> 3, 4, 5, 3, 4, 5, 3, 4, 5, 3, 4, 5,~
## $ is_blank            <lgl> FALSE, FALSE, FALSE, FALSE, FALSE, ~
## $ content             <chr> "59597300", "30420100", "29177200",~
## $ data_type           <chr> "numeric", "numeric", "numeric", "n~
## $ error               <chr> NA, NA, NA, NA, NA, NA, NA, NA, NA,~
## $ logical             <lgl> NA, NA, NA, NA, NA, NA, NA, NA, NA,~
## $ numeric             <dbl> 59597300, 30420100, 29177200, 56489~
## $ date                <dttm> NA, NA, NA, NA, NA, NA, NA, NA, NA~
## $ character           <chr> NA, NA, NA, NA, NA, NA, NA, NA, NA,~
## $ character_formatted <list> <NULL>, <NULL>, <NULL>, <NULL>, <N~
## $ formula             <chr> NA, NA, NA, NA, NA, NA, NA, NA, NA,~
## $ is_array            <lgl> FALSE, FALSE, FALSE, FALSE, FALSE, ~
## $ formula_ref         <chr> NA, NA, NA, NA, NA, NA, NA, NA, NA,~
## $ formula_group       <int> NA, NA, NA, NA, NA, NA, NA, NA, NA,~
## $ comment             <chr> NA, NA, NA, NA, NA, NA, NA, NA, NA,~
## $ height              <dbl> 15.6, 15.6, 15.6, 15.6, 15.6, 15.6,~
## $ width               <dbl> 14.71, 12.43, 12.43, 14.71, 12.43, ~
## $ row_outline_level   <dbl> 1, 1, 1, 1, 1, 1, 1, 1, 1, 1, 1, 1,~
## $ col_outline_level   <dbl> 1, 1, 1, 1, 1, 1, 1, 1, 1, 1, 1, 1,~
## $ style_format        <chr> "Normal", "Normal", "Normal", "Norm~
## $ local_format_id     <int> 33, 33, 33, 33, 33, 33, 33, 33, 33,~
## $ field0              <chr> "England and Wales", "England and W~
## $ field1              <chr> NA, NA, NA, "England", "England", "~
## $ field2              <chr> NA, NA, NA, NA, NA, NA, "North East~
## $ field3              <chr> NA, NA, NA, NA, NA, NA, NA, NA, NA,~
## $ field4              <chr> NA, NA, NA, NA, NA, NA, NA, NA, NA,~
## $ gender              <chr> "All persons", "Females", "Males", ~
```

We can also look at the first few rows of the variables we are interested in, with a combination of `select()` and `slice_head()`.

```
uk_census_behead |>
  select(field0:field4, gender, numeric) |>
  slice_head(n = 10)
```

```
## # A tibble: 10 x 7
##    field0    field1 field2 field3 field4 gender numeric
##    <chr>     <chr>  <chr>  <chr>  <chr>  <chr>    <dbl>
##  1 England ~ <NA>   <NA>   <NA>   <NA>   All p~  5.96e7
##  2 England ~ <NA>   <NA>   <NA>   <NA>   Femal~  3.04e7
##  3 England ~ <NA>   <NA>   <NA>   <NA>   Males   2.92e7
##  4 England ~ Engla~ <NA>   <NA>   <NA>   All p~  5.65e7
##  5 England ~ Engla~ <NA>   <NA>   <NA>   Femal~  2.88e7
##  6 England ~ Engla~ <NA>   <NA>   <NA>   Males   2.77e7
##  7 England ~ Engla~ North~ <NA>   <NA>   All p~  2.65e6
##  8 England ~ Engla~ North~ <NA>   <NA>   Femal~  1.35e6
##  9 England ~ Engla~ North~ <NA>   <NA>   Males   1.29e6
## 10 England ~ Engla~ North~ Count~ <NA>   All p~  5.22e5
```

So far, so good.

And the bottom 10 rows, using `slice_tail()`.

```
uk_census_behead |>
  select(field0:field4, gender, numeric) |>
  slice_tail(n = 10)
```

```
## # A tibble: 10 x 7
##    field0    field1 field2 field3 field4 gender numeric
##    <chr>     <chr>  <chr>  <chr>  <chr>  <chr>    <dbl>
##  1 England ~ Wales  South~ Somer~ Blaen~ Males    32800
##  2 England ~ Wales  South~ Somer~ Torfa~ All p~   92300
##  3 England ~ Wales  South~ Somer~ Torfa~ Femal~   47400
##  4 England ~ Wales  South~ Somer~ Torfa~ Males    44900
##  5 England ~ Wales  South~ Somer~ Monmo~ All p~   93000
##  6 England ~ Wales  South~ Somer~ Monmo~ Femal~   47400
##  7 England ~ Wales  South~ Somer~ Monmo~ Males    45600
##  8 England ~ Wales  South~ Somer~ Newpo~ All p~  159600
##  9 England ~ Wales  South~ Somer~ Newpo~ Femal~   81200
## 10 England ~ Wales  South~ Somer~ Newpo~ Males    78400
```

Wait! Didn't we learn that Wales is not divided into regions, and that there are no divisions of the Unitary Authority level? Furthermore, you may also already know that Somerset is in England, not Wales. What's gone wrong?

A check of the tail of the data (above) reveals a problem with the structure of the Excel file, which (mostly) reflects the nature of the hierarchy. This is not a problem with the {tidyr} and {unpivotr} functions! Those functions are doing things exactly correctly. The values are being pulled from the value that is found at "left-up".

The first thing that catches my eye is that `field2`—the Region—should be NA for Wales, but it has been assigned to the South West, which is the last English Region in the data.

More problematically, the Welsh Unitary Authorities (W06) should be in `field3`, at the same level as the English Unitary Authorities (E06), but instead have been assigned to `field4`. This has led to Somerset, in England, to be assigned to the column where we expect to see the names of the Welsh Unitary Authorities.

Let's take a look at the details of the first row of the Welsh Unitary Authorities. First, in the `uk_census` object, we see that the value of `local_format_id` for the name ("Isle of Anglesey") is 60.

```
uk_census |>
  dplyr::filter((row == 361)) |>
  select(address, local_format_id, data_type, character, numeric)
```

```
## # A tibble: 5 x 5
##   address local_format_id data_type character   numeric
##   <chr>             <int> <chr>     <chr>         <dbl>
## 1 A361                 11 character W06000001        NA
## 2 B361                 60 character Isle of An~      NA
## 3 C361                 33 numeric   <NA>          68900
## 4 D361                 33 numeric   <NA>          35200
## 5 E361                 33 numeric   <NA>          33700
```

We can then use this to examine the details of the indent formatting, which we captured in the `formats` object.

```
uk_census_formats$local$alignment$indent[60]
```

```
## [1] 4
```

There are 4 indents. You will recall that the `behead_if()` function was looking left and up. In the case of the Welsh entries, the function has looked left and up until it found an entry that has fewer indents, which happens to be in England, not Wales.

This problem has arisen at the transition from one area to the next. Did this also happen in the England rows, for example where the North West region row appears? As we can see below, as similar problem has occurred: the value "North West" is correctly assigned to `field2`, but `field3` has been populated with the first value *above* North West where there is a valid value in `field3`.

```
uk_census_behead |>
  select(address, content, field0, field1, field2, field3, field4) |>
  filter(field2 == "North West")
```

```
## # A tibble: 132 x 7
##    address content field0   field1 field2 field3 field4
##    <chr>   <chr>   <chr>    <chr>  <chr>  <chr>  <chr>
##  1 C24     7417300 England~ Engla~ North~ Tyne ~ <NA>
##  2 D24     3777200 England~ Engla~ North~ Tyne ~ <NA>
##  3 E24     3640100 England~ Engla~ North~ Tyne ~ <NA>
##  4 C25     154800  England~ Engla~ North~ Black~ <NA>
##  5 D25     78100   England~ Engla~ North~ Black~ <NA>
##  6 E25     76800   England~ Engla~ North~ Black~ <NA>
##  7 C26     141100  England~ Engla~ North~ Black~ <NA>
##  8 D26     71400   England~ Engla~ North~ Black~ <NA>
##  9 E26     69700   England~ Engla~ North~ Black~ <NA>
## 10 C27     398800  England~ Engla~ North~ Chesh~ <NA>
## # i 122 more rows
```

Now that we have diagnosed the problem, what solution can we apply?

One solution is to repopulate each level's `field` variable by stepping up a level and comparing that to the value three rows earlier in the data[5]. (We look at the previous rows using the `dplyr::lag()` function. It has to be three rows because our beheading function has made separate rows for "All persons", "Female", and "Male".)

If the prior field is not the same, it gets replaced with an "NA", but if it's the same, then the current value is retained. In the first `mutate()` function, `field1` is replaced if `field0` in the current spot is *not* the same as the `field0` three rows prior.

[5] Thanks to Duncan Garmonsway, the author of the {tidyxl} and {untidyr} packages, for the path to this solution.

```
uk_census_behead |>
  select(row, col, starts_with("field")) |>
  arrange(row, col) |>
  mutate(field1 = if_else(field0 == lag(field0, n = 3), field1, NA)) |>
  mutate(field2 = if_else(field1 == lag(field1, n = 3), field2, NA)) |>
  mutate(field3 = if_else(field2 == lag(field2, n = 3), field3, NA)) |>
  mutate(field4 = if_else(field3 == lag(field3, n = 3), field4, NA))
```

```
## # A tibble: 1,125 x 7
##      row   col field0       field1 field2 field3 field4
##    <int> <int> <chr>        <chr>  <chr>  <chr>  <chr>
##  1     8     3 England and~ <NA>   <NA>   <NA>   <NA>
##  2     8     4 England and~ <NA>   <NA>   <NA>   <NA>
##  3     8     5 England and~ <NA>   <NA>   <NA>   <NA>
##  4     9     3 England and~ Engla~ <NA>   <NA>   <NA>
##  5     9     4 England and~ Engla~ <NA>   <NA>   <NA>
##  6     9     5 England and~ Engla~ <NA>   <NA>   <NA>
##  7    10     3 England and~ Engla~ North~ <NA>   <NA>
##  8    10     4 England and~ Engla~ North~ <NA>   <NA>
##  9    10     5 England and~ Engla~ North~ <NA>   <NA>
## 10    11     3 England and~ Engla~ North~ Count~ <NA>
## # i 1,115 more rows
```

Unfortunately, this solution does not repair the mis-assignment of the UAs in Wales.

In the code below, the values in `field1` are evaluated. For those with population data from Wales, the `field2` and `field3` values are replaced with an NA.

```
uk_census_behead_revised <- uk_census_behead |>
  mutate(field2 = case_when(
    field1 == "Wales" ~ NA_character_,
    TRUE ~ field2
  )) |>
  mutate(field3 = case_when(
    field1 == "Wales" ~ NA_character_,
    TRUE ~ field3
  )) |>
  select(field0:field4, gender, numeric)

uk_census_behead_revised |>
  tail()
```

```
## # A tibble: 6 x 7
##   field0     field1 field2 field3 field4 gender numeric
##   <chr>      <chr>  <chr>  <chr>  <chr>  <chr>    <dbl>
## 1 England a~ Wales  <NA>   <NA>   Monmo~ All p~   93000
## 2 England a~ Wales  <NA>   <NA>   Monmo~ Femal~   47400
## 3 England a~ Wales  <NA>   <NA>   Monmo~ Males    45600
## 4 England a~ Wales  <NA>   <NA>   Newpo~ All p~  159600
## 5 England a~ Wales  <NA>   <NA>   Newpo~ Femal~   81200
## 6 England a~ Wales  <NA>   <NA>   Newpo~ Males    78400
```

In many uses of this data, we will want to remove the subtotal rows. In the code below, a `case_when()` function is applied to create a new variable that contains a value "subtotal" for the rows that are disaggregated at the next level (for example, the cities within a Metro county, or the counties within a region).

The first step is to create a new dataframe `field3_split` that identifies all of the areas at the county level that have a dissaggregation. This is done by first filtering by `field4`, where there is a valid value (identified as not an "NA"), and creating a list of the distinct values in `field3`. That list is then used as part of the evaluation within the `case_when()` to assign the value "subtotal" to the new variable `subtotal_row`.

```
field3_split <- uk_census_behead_revised |>
  filter(!is.na(field4)) |>
  distinct(field3)

field3_split
```

```
## # A tibble: 33 x 1
##    field3
##    <chr>
##  1 Tyne and Wear (Met County)
##  2 Cumbria
##  3 Greater Manchester (Met County)
##  4 Lancashire
##  5 Merseyside (Met County)
##  6 North Yorkshire
##  7 South Yorkshire (Met County)
##  8 West Yorkshire (Met County)
##  9 Derbyshire
## 10 Leicestershire
## # i 23 more rows
```

```
uk_census_behead_revised |>
  mutate(subtotal_row = case_when(
    field3 %in% field3_split$field3 & is.na(field4) ~ "subtotal",
    TRUE ~ NA_character_
    )) |>
  relocate(subtotal_row)
```

```
## # A tibble: 1,125 x 8
##    subtotal_row field0       field1 field2 field3 field4
##    <chr>        <chr>        <chr>  <chr>  <chr>  <chr>
##  1 subtotal     England an~ <NA>   <NA>   <NA>   <NA>
##  2 subtotal     England an~ <NA>   <NA>   <NA>   <NA>
##  3 subtotal     England an~ <NA>   <NA>   <NA>   <NA>
##  4 subtotal     England an~ Engla~ <NA>   <NA>   <NA>
##  5 subtotal     England an~ Engla~ <NA>   <NA>   <NA>
##  6 subtotal     England an~ Engla~ <NA>   <NA>   <NA>
##  7 subtotal     England an~ Engla~ North~ <NA>   <NA>
##  8 subtotal     England an~ Engla~ North~ <NA>   <NA>
##  9 subtotal     England an~ Engla~ North~ <NA>   <NA>
## 10 <NA>         England an~ Engla~ North~ Count~ <NA>
## # i 1,115 more rows
## # i 2 more variables: gender <chr>, numeric <dbl>
```

Approach 3 - build a concordance table

If we need a more flexible and re-usable approach, we can consider the option of creating a separate concordance table (also known as a crosswalk table). A table of this sort contains the distinct values of key variables and their corresponding values in another typology or schema. Once a table like this is built, it can be joined to our data, as well as to any other table we might need to use in the future.

For the UK Census data, we will create two separate concordance tables. We will save these tables as CSV files, so they are readily available for other instances when their use will make filtering and summarizing the data more efficient.

For the first table, we will build a table with each type of administrative region and its associated three-digit code. This is done programatically in R, using the {tibble} package (Müller and Wickham, 2021). We will use the `tribble()` function, which transposes the layout so it resembles the final tabular structure.

In addition to the area code prefix and the name of the administrative region type, the level in the hierarchy is also encoded in this table.

```
uk_region_table <- tribble(
  # header row
  ~"area_code_3", ~"category", ~"geo_level",
  "K04", "England & Wales", 0,
  "E92", "England", 1,
  "E12", "Region", 2,
  "E09", "London Borough", 3,
  "E10", "County",  3,
  "E07", "Non-Metropolitan District", 4,
  "E06", "Unitary Authority", 3,
  "E11", "Metropolitan County", 3,
  "E08", "Metropolitan District", 4,
  "W92", "Wales", 1,
  "W06", "Unitary Authority", 3
)

uk_region_table
```

```
## # A tibble: 11 x 3
##    area_code_3 category                  geo_level
##    <chr>       <chr>                         <dbl>
##  1 K04         England & Wales                   0
##  2 E92         England                           1
##  3 E12         Region                            2
##  4 E09         London Borough                    3
##  5 E10         County                            3
##  6 E07         Non-Metropolitan District         4
##  7 E06         Unitary Authority                 3
##  8 E11         Metropolitan County               3
##  9 E08         Metropolitan District             4
## 10 W92         Wales                             1
## 11 W06         Unitary Authority                 3
```

This table can be saved as a CSV file for future reference:

```
write_csv(uk_region_table, "data_output/uk_region_table.csv")
```

Our next step is to use a `mutate()` function to extract the unique values in the area code variable in the dataframe `uk_census_pop`, and then use that new variable to join the dataframe to the `uk_region_table` created in the previous chunk.

```
uk_census_pop_geo <- uk_census_pop |>
#  select(area_code, area_name) |>
  mutate(area_code_3 = str_sub(uk_census_pop$area_code, 1, 3)) |>
  # join with classification table
  left_join(
    uk_region_table,
    by = c("area_code_3" = "area_code_3")
    )

glimpse(uk_census_pop_geo, width = 65)
```

```
## Rows: 375
## Columns: 8
## $ area_code   <chr> "K04000001", "E92000001", "E12000001", "E06~
## $ area_name   <chr> "England and Wales", "England", "North East~
## $ all_persons <dbl> 59597300, 56489800, 2647100, 522100, 107800~
## $ females     <dbl> 30420100, 28833500, 1353800, 266800, 55100,~
## $ males       <dbl> 29177200, 27656300, 1293300, 255300, 52700,~
## $ area_code_3 <chr> "K04", "E92", "E12", "E06", "E06", "E06", "~
## $ category    <chr> "England & Wales", "England", "Region", "Un~
## $ geo_level   <dbl> 0, 1, 2, 3, 3, 3, 3, 3, 3, 3, 3, 4, 4, 4, 4~
```

Because the level is now encoded in the relevant rows of the dataframe, it is possible to filter on the level variable to calculate the total population for England and Wales.

```
uk_census_pop_geo |>
  filter(geo_level == 3) |>
  summarize(total_population = sum(all_persons))
```

```
## # A tibble: 1 x 1
##   total_population
##              <dbl>
## 1         59597300
```

A more complex concordance table can be derived from the beheaded table we created earlier. The two steps in this process are:

- Select the field variables and then create a table with only the distinct values across all the rows,

- Create a new variable "area_name" with the lowest-level value, i.e. the values in the source table column B.

```
uk_census_geo_code <- uk_census_behead_revised |>
  # select field variables
  select(contains("field")) |>
  distinct() |>
  # create variable with lowest-level value
  mutate(area_name = case_when(
    !is.na(field4) ~ field4,
    !is.na(field3) ~ field3,
    !is.na(field2) ~ field2,
    !is.na(field1) ~ field1,
    !is.na(field0) ~ field0,
    TRUE ~ NA_character_
  ))

uk_census_geo_code
```

```
## # A tibble: 375 x 6
##    field0          field1 field2 field3 field4 area_name
##    <chr>           <chr>  <chr>  <chr>  <chr>  <chr>
##  1 England and W~ <NA>   <NA>   <NA>   <NA>   England ~
##  2 England and W~ Engla~ <NA>   <NA>   <NA>   England
##  3 England and W~ Engla~ North~ <NA>   <NA>   North Ea~
##  4 England and W~ Engla~ North~ Count~ <NA>   County D~
##  5 England and W~ Engla~ North~ Darli~ <NA>   Darlingt~
##  6 England and W~ Engla~ North~ Hartl~ <NA>   Hartlepo~
##  7 England and W~ Engla~ North~ Middl~ <NA>   Middlesb~
##  8 England and W~ Engla~ North~ North~ <NA>   Northumb~
##  9 England and W~ Engla~ North~ Redca~ <NA>   Redcar a~
## 10 England and W~ Engla~ North~ Stock~ <NA>   Stockton~
## # i 365 more rows
```

The third step is to join the area code variable from the `uk_census_pop` table.

```
uk_census_geo_code <- uk_census_geo_code |>
  left_join(
    select(uk_census_pop, area_name, area_code),
    by = "area_name"
    ) |>
```

```
  relocate(area_name, area_code)

uk_census_geo_code
```

```
## # A tibble: 375 x 7
##    area_name     area_code field0 field1 field2 field3
##    <chr>         <chr>     <chr>  <chr>  <chr>  <chr>
##  1 England and W~ K04000001 Engla~ <NA>   <NA>   <NA>
##  2 England        E92000001 Engla~ Engla~ <NA>   <NA>
##  3 North East     E12000001 Engla~ Engla~ North~ <NA>
##  4 County Durham  E06000047 Engla~ Engla~ North~ Count~
##  5 Darlington     E06000005 Engla~ Engla~ North~ Darli~
##  6 Hartlepool     E06000001 Engla~ Engla~ North~ Hartl~
##  7 Middlesbrough  E06000002 Engla~ Engla~ North~ Middl~
##  8 Northumberland E06000057 Engla~ Engla~ North~ North~
##  9 Redcar and Cl~ E06000003 Engla~ Engla~ North~ Redca~
## 10 Stockton-on-T~ E06000004 Engla~ Engla~ North~ Stock~
## # i 365 more rows
## # i 1 more variable: field4 <chr>
```

To this table the geographic level values are added by joining the `uk_region_table` created earlier.

```
uk_census_geo_code <- uk_census_geo_code |>
#  select(area_code, area_name) |>
  mutate(area_code_3 = str_sub(uk_census_pop$area_code, 1, 3)) |>
  # join with classification table
  left_join(
    uk_region_table,
    by = c("area_code_3" = "area_code_3")
    )

uk_census_geo_code
```

```
## # A tibble: 375 x 10
##    area_name     area_code field0 field1 field2 field3
##    <chr>         <chr>     <chr>  <chr>  <chr>  <chr>
##  1 England and W~ K04000001 Engla~ <NA>   <NA>   <NA>
##  2 England        E92000001 Engla~ Engla~ <NA>   <NA>
##  3 North East     E12000001 Engla~ Engla~ North~ <NA>
```

```
##  4 County Durham  E06000047 Engla~ Engla~ North~ Count~
##  5 Darlington     E06000005 Engla~ Engla~ North~ Darli~
##  6 Hartlepool     E06000001 Engla~ Engla~ North~ Hartl~
##  7 Middlesbrough  E06000002 Engla~ Engla~ North~ Middl~
##  8 Northumberland E06000057 Engla~ Engla~ North~ North~
##  9 Redcar and Cl~ E06000003 Engla~ Engla~ North~ Redca~
## 10 Stockton-on-T~ E06000004 Engla~ Engla~ North~ Stock~
## # i 365 more rows
## # i 4 more variables: field4 <chr>, area_code_3 <chr>,
## #   category <chr>, geo_level <dbl>
```

Now we have a comprehensive table, with all 375 of the administrative geography areas, including the name, area code, and all of the details of the hierarchy.

This table could also be joined to the contents of a population sheet, to provide additional detail.

Let's imagine your assignment is to determine which region at the UA level has the highest proportion of people aged 90 or older. The code below reads the contents of sheet "P02" in the census population file, which has age detail of the population in each area. By appending the contents of the concordance table, a detailed disaggregation without double counting is possible.

In addition to changing the sheet reference, the code also changes the range to reflect the differences in the contents.

```
uk_census_pop_90 <- readxl::read_excel(
  dpjr::dpjr_data("census2021firstresultsenglandwales1.xlsx"),
  sheet = "P02",
  range = "A8:V383"
) |>
  janitor::clean_names() |>
  dplyr::rename(area_code = area_code_note_2) |>
  # remove "_note_12" from population variables
  dplyr::rename_with(~str_remove(., "_note_12"))

head(uk_census_pop_90)
```

```
## # A tibble: 6 x 22
##   area_code area_name         all_persons
##   <chr>     <chr>                   <dbl>
## 1 K04000001 England and Wales    59597300
## 2 E92000001 England              56489800
## 3 E12000001 North East            2647100
```

```
## 4 E06000047 County Durham              522100
## 5 E06000005 Darlington                 107800
## 6 E06000001 Hartlepool                  92300
## # i 19 more variables: aged_4_years_and_under <dbl>,
## #   aged_5_to_9_years <dbl>,
## #   aged_10_to_14_years <dbl>,
## #   aged_15_to_19_years <dbl>,
## #   aged_20_to_24_years <dbl>,
## #   aged_25_to_29_years <dbl>,
## #   aged_30_to_34_years <dbl>, ...
```

The `filter()` step below does the following:

- selects the area code and the population categories of interest,
- joins the detailed geographic descriptions,
- filters the level 3 values, but this leaves those UA regions that are further disaggregated into the city category.

```
uk_census_pop_90 <- uk_census_pop_90 |>
  select(area_code, all_persons, aged_90_years_and_over) |>
  full_join(uk_census_geo_code, by = "area_code") |>
  # filter so that only UAs remain
  filter(geo_level == 3)
```

With this table, the proportion of persons aged 90 or older can be calculated and the table sorted in descending order:

```
uk_census_pop_90 |>
  mutate(
    pct_90_plus =
      round((aged_90_years_and_over / all_persons * 100), 1)
    ) |>
  arrange(desc(pct_90_plus)) |>
  select(area_name, aged_90_years_and_over, pct_90_plus) |>
  slice_head(n = 10)
```

```
## # A tibble: 10 x 3
##    area_name           aged_90_years_and_over pct_90_plus
##    <chr>                                <dbl>       <dbl>
```

```
## 1 Dorset [note 10]                         6300       1.7
## 2 East Sussex                              8100       1.5
## 3 Conwy                                    1700       1.5
## 4 Isle of Wight                            2000       1.4
## 5 Torbay                                   2000       1.4
## 6 Devon                                   11200       1.4
## 7 Herefordshire, C~                        2400       1.3
## 8 West Sussex                             11900       1.3
## 9 Bournemouth, Chr~                        5400       1.3
## 10 North Somerset                          2800       1.3
```

7

Importing data: statistical software

In this chapter:

- Importing survey data from statistical software packages that include labelled data formats, with an emphasis on SPSS
- Importing survey data from an Excel file, including the data dictionary created by survey software, and creating labels from the data dictionary.
- Importing plain-text data and applying syntax from statistical software packages to apply labels and output as statistical software data formats, with an emphasis on SPSS.

7.1 Statistical software

If you are one or work with statisticians, economists, sociologists, survey practitioners, and many others, you will find yourself encountering data files created using the software packages SAS, SPSS, and Stata. These tools have been around a long time (SAS and SPSS trace their history back to the late 1960s, and Stata was created in 1985), and over the years they have evolved, demonstrated their robustness, and have become common in many academic, corporate, and government settings.

A feature of these programs is that the variables and the associated values can be *labelled*. If you're familiar with **R**, value labelling is conceptually similar to the use of factor labels. These labels can carry a great deal of detail to which you might otherwise not have easy access.

The **R** package {haven} (Wickham and Miller, 2021) provides the functionality to read these three types of files[1]. There are other packages related to {haven} that we will also use.

```
library(haven)
```

[1] The reference page for {haven} is here: https://haven.tidyverse.org/index.html

In the following examples, we will use versions of the data in the {palmerpenguins} data (Horst et al., 2022; Horst, 2020). We will work with SPSS formatted files; the approach is similar for SAS and Stata formats, and only differs in the details. For example, the `haven::read_()` functions vary only with the extension; a Stata data file has the extension ".dta", so the read function is `read_dta()`:

```
haven::read_dta(dpjr::dpjr_data("penguins.dta"))
```

The equivalent for a SAS file is as follows:

```
haven::read_sas(dpjr::dpjr_data("penguins.sas"))
```

Here's the code to read that same data file in SPSS's ".sav" format and assigns it to an R object `penguins_sav`:

```
penguins_sav <- haven::read_sav(dpjr::dpjr_data("penguins.sav"))

penguins_sav
```

```
## # A tibble: 344 x 8
##    species     island        bill_length_mm bill_depth_mm
##    <dbl+lbl>   <dbl+lbl>              <dbl>         <dbl>
##  1 1 [Adelie] 3 [Torgerse~             39.1          18.7
##  2 1 [Adelie] 3 [Torgerse~             39.5          17.4
##  3 1 [Adelie] 3 [Torgerse~             40.3          18
##  4 1 [Adelie] 3 [Torgerse~             NA            NA
##  5 1 [Adelie] 3 [Torgerse~             36.7          19.3
##  6 1 [Adelie] 3 [Torgerse~             39.3          20.6
##  7 1 [Adelie] 3 [Torgerse~             38.9          17.8
##  8 1 [Adelie] 3 [Torgerse~             39.2          19.6
##  9 1 [Adelie] 3 [Torgerse~             34.1          18.1
## 10 1 [Adelie] 3 [Torgerse~             42            20.2
## # i 334 more rows
## # i 4 more variables: flipper_length_mm <dbl>,
## #   body_mass_g <dbl>, sex <dbl+lbl>, year <dbl>
```

You will notice that the values of the variables in the `.sav` version of the `penguins` dataframe are different than what was read in from the `.csv` version. All of the variables, including those that were character strings, are numeric.

For example in the variable "species" the values are a column of "1"s, "2"s, and "3"s instead of being the names of the species.

Note also that in the variable descriptions, it includes `<S3: haven_labelled>` for those that bring the SPSS labels with them.

7.1.1 {labelled} -

Once we have read our SAS, SPSS, or Stata data into R using one of the {haven} functions, the package {labelled}(Larmarange, 2022) gives us a range of powerful tools for working with the variable labels, value labels, and defined missing values.

```
library(labelled)
```

The first function we will use is `look_for()`, which returns the labels, column types, and values of the variables in our data.

```
look_for(penguins_sav)
```

```
##  pos variable          label col_type missing
##  1   species           —     dbl+lbl  0
##
##
##  2   island            —     dbl+lbl  0
##
##
##  3   bill_length_mm    —     dbl      2
##  4   bill_depth_mm     —     dbl      2
##  5   flipper_length_mm —     dbl      2
##  6   body_mass_g       —     dbl      2
##  7   sex               —     dbl+lbl  11
##
##  8   year              —     dbl      0
##  values
##  [1] Adelie
##  [2] Chinstrap
##  [3] Gentoo
##  [1] Biscoe
##  [2] Dream
##  [3] Torgersen
##
##
```

```
##
##
##   [1] female
##   [2] male
##
```

In this view, we can spot the species names in the "values" column. Also notice that the "col_type" column shows "dbl+lbl": double and label. In this dataframe, no labels have been defined.

The function `var_label()` allows us to add a descriptive label to a variable. In the next code chunk, we add a label to the `species` variable:

```
var_label(penguins_sav$species) <- "Penguin species"
```

We can see that the label is now incorporated into the dataframe using the `look_for()` function:

```
look_for(penguins_sav)
```

```
##  pos variable          label     col_type missing
##  1   species           Penguin ~ dbl+lbl  0
##
##
##  2   island            —         dbl+lbl  0
##
##
##  3   bill_length_mm    —         dbl      2
##  4   bill_depth_mm     —         dbl      2
##  5   flipper_length_mm —         dbl      2
##  6   body_mass_g       —         dbl      2
##  7   sex               —         dbl+lbl  11
##
##  8   year              —         dbl      0
##  values
##  [1] Adel~
##  [2] Chin~
##  [3] Gent~
##  [1] Bisc~
##  [2] Dream
##  [3] Torg~
##
##
```

```
##
##
## [1] fema~
## [2] male
##
```

It's also possible to add multiple labels. Note that in this example, a label is added to both "dbl+lbl" and "dbl" types.

```
var_label(penguins_sav) <- list(island = "Island of record",
                                 body_mass_g = "Weight, in grams")
```

```
look_for(penguins_sav)
```

```
## pos variable          label      col_type missing
## 1   species           Penguin ~ dbl+lbl  0
##
##
## 2   island            Island o~ dbl+lbl  0
##
##
## 3   bill_length_mm    —         dbl      2
## 4   bill_depth_mm     —         dbl      2
## 5   flipper_length_mm —         dbl      2
## 6   body_mass_g       Weight, ~ dbl      2
## 7   sex               —         dbl+lbl  11
##
## 8   year              —         dbl      0
## values
## [1] Adel~
## [2] Chin~
## [3] Gent~
## [1] Bisc~
## [2] Dream
## [3] Torg~
##
##
##
##
## [1] fema~
## [2] male
##
```

7.1.1.1 Using labels

The labels in an dataframe created by {haven} can be converted to a factor type, using the `unlabelled()` function.

If we were to tally the number of penguins in each species using the penguins dataframe that originated from the SPSS file, we would see this:

```
penguins_sav |>
  group_by(species) |>
  tally()
```

```
## # A tibble: 3 x 2
##   species             n
##   <dbl+lbl>       <int>
## 1 1 [Adelie]        152
## 2 2 [Chinstrap]      68
## 3 3 [Gentoo]        124
```

In this version, the "species" variable retains the numeric values.

By applying the `unlabelled()` function, the "haven_labelled" type variables are transformed into factor types, with the labels (that is, the species names) now the shown in our summary table:

```
penguins_sav |>
  unlabelled() |>
  group_by(species) |>
  tally()
```

```
## # A tibble: 3 x 2
##   species         n
##   <fct>       <int>
## 1 Adelie        152
## 2 Chinstrap      68
## 3 Gentoo        124
```

An alternative approach is to the explicitly change the variable type, with a `to_factor` transformation inside one of the mutate functions, `mutate()` and `mutate_if()`. In this first example, the variable `species` is mutated from labelled to factor type.

```
penguins_sav |>
  mutate(species = as_factor(species)) |>
  group_by(species) |>
  tally()
```

```
## # A tibble: 3 x 2
##   species       n
##   <fct>     <int>
## 1 Adelie      152
## 2 Chinstrap    68
## 3 Gentoo      124
```

In this second variant, the `mutate_if(is.labelled)` function transforms all of the labelled variables.

```
penguins_sav |>
  mutate_if(is.labelled, to_factor) |>
  group_by(species, island) |>
  tally()
```

```
## # A tibble: 5 x 3
## # Groups:   species [3]
##   species   island        n
##   <fct>     <fct>     <int>
## 1 Adelie    Biscoe       44
## 2 Adelie    Dream        56
## 3 Adelie    Torgersen    52
## 4 Chinstrap Dream        68
## 5 Gentoo    Biscoe      124
```

7.1.2 Reading an SPSS survey file: "Video"

The functionality of a labelled dataframe is very beneficial when working with survey data. The dataframe can contain the variable codes as captured in the form, the values associated with those values, as well as the text of the question in the label.

The file used in this example comes from the University of Sheffield Mathematics and Statistics Help's "Datasets for Teaching"[2]

[2]https://www.sheffield.ac.uk/mash/statistics/datasets

This dataset was collected by Scott Smith (University of Sheffield) to evaluate the use of best method for informing the public about a certain medical condition. There were three videos (New general video A, new medical profession video B, the old video C and a demonstration using props D). He wanted to see if the new methods were more popular so collected data using mostly Likert style questions about a range of things such as understanding and general impressions. This reduced dataset contains some of those questions and 4 scale scores created from summing 5 ordinal questions to give a scale score.

If you were working in SPSS, when you open the file, you would see the data like this:

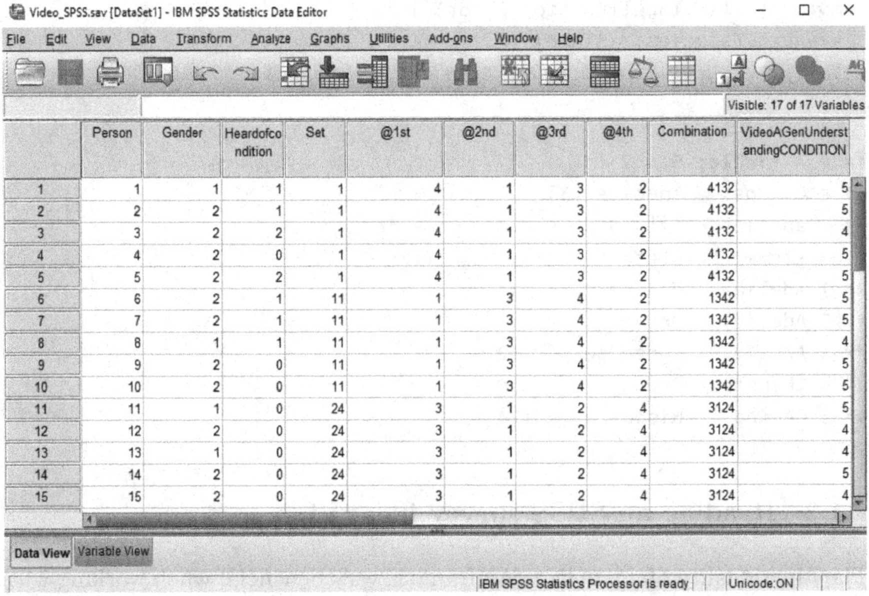

Video_SPSS.sav [DataSet1] - IBM SPSS Statistics Data Editor

File Edit View Data Transform Analyze Graphs Utilities Add-ons Window Help

Visible: 17 of 17 Variables

	Person	Gender	Heardofcondition	Set	@1st	@2nd	@3rd	@4th	Combination	VideoAGenUnderstandingCONDITION
1	1	1	1	1	4	1	3	2	4132	5
2	2	2	1	1	4	1	3	2	4132	5
3	3	2	2	1	4	1	3	2	4132	4
4	4	2	0	1	4	1	3	2	4132	5
5	5	2	2	1	4	1	3	2	4132	5
6	6	2	1	11	1	3	4	2	1342	5
7	7	2	1	11	1	3	4	2	1342	5
8	8	1	1	11	1	3	4	2	1342	4
9	9	2	0	11	1	3	4	2	1342	5
10	10	2	0	11	1	3	4	2	1342	5
11	11	1	0	24	3	1	2	4	3124	5
12	12	2	0	24	3	1	2	4	3124	4
13	13	1	0	24	3	1	2	4	3124	4
14	14	2	0	24	3	1	2	4	3124	5
15	15	2	0	24	3	1	2	4	3124	4

Data View Variable View

IBM SPSS Statistics Processor is ready Unicode:ON

FIGURE 7.1 *SPSS data view.*

SPSS also gives you the option to view the variables:

In this view, we can see both the variable names and the variable labels (in this case, the precise wording of the survey question).

And we can also drill deeper, and see the value labels:

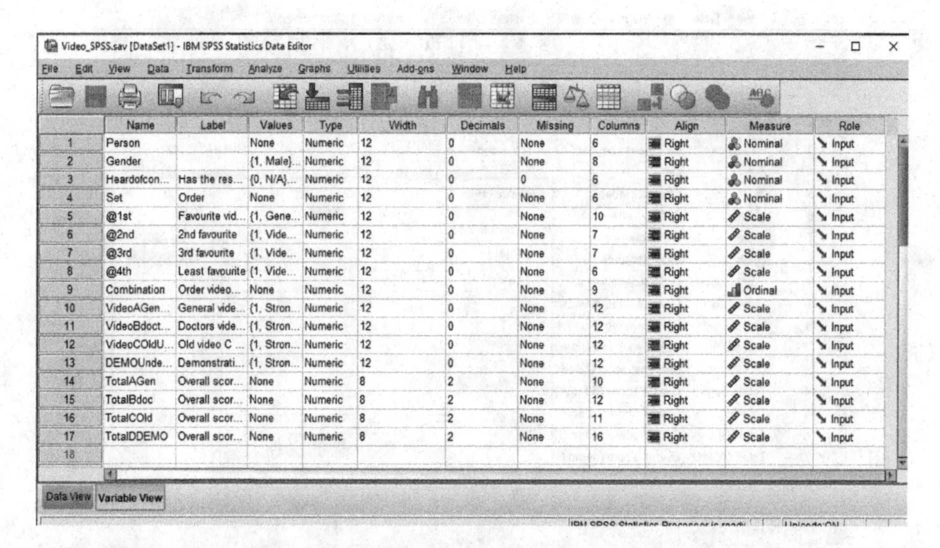

Video_SPSS.sav [DataSet1] - IBM SPSS Statistics Data Editor

File Edit View Data Transform Analyze Graphs Utilities Add-ons Window Help

	Name	Label	Values	Type	Width	Decimals	Missing	Columns	Align	Measure	Role
1	Person		None	Numeric	12	0	None	6	Right	Nominal	Input
2	Gender		{1, Male}...	Numeric	12	0	None	8	Right	Nominal	Input
3	Heardofcon...	Has the res...	{0, N/A}...	Numeric	12	0	0	6	Right	Nominal	Input
4	Set	Order	None	Numeric	12	0	None	6	Right	Nominal	Input
5	@1st	Favourite vid...	{1, Gene...	Numeric	12	0	None	10	Right	Scale	Input
6	@2nd	2nd favourite	{1, Vide...	Numeric	12	0	None	7	Right	Scale	Input
7	@3rd	3rd favourite	{1, Vide...	Numeric	12	0	None	7	Right	Scale	Input
8	@4th	Least favourite	{1, Vide...	Numeric	12	0	None	6	Right	Scale	Input
9	Combination	Order video...	None	Numeric	12	0	None	9	Right	Ordinal	Input
10	VideoAGen...	General vide...	{1, Stron...	Numeric	12	0	None	12	Right	Scale	Input
11	VideoBdoct...	Doctors vide...	{1, Stron...	Numeric	12	0	None	12	Right	Scale	Input
12	VideoCOldU...	Old video C ...	{1, Stron...	Numeric	12	0	None	12	Right	Scale	Input
13	DEMOUnde...	Demonstrati...	{1, Stron...	Numeric	12	0	None	12	Right	Scale	Input
14	TotalAGen	Overall scor...	None	Numeric	8	2	None	10	Right	Scale	Input
15	TotalBdoc	Overall scor...	None	Numeric	8	2	None	12	Right	Scale	Input
16	TotalCOld	Overall scor...	None	Numeric	8	2	None	11	Right	Scale	Input
17	TotalDDEMO	Overall scor...	None	Numeric	8	2	None	16	Right	Scale	Input
18											

Data View Variable View

FIGURE 7.2 *SPSS variable view.*

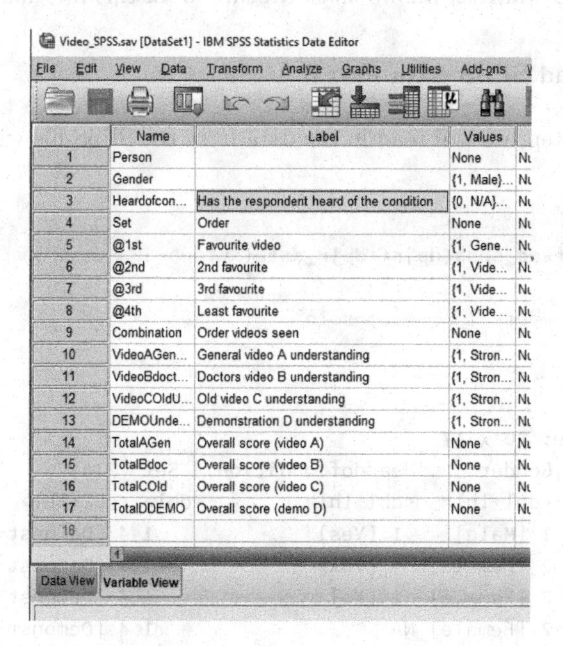

Video_SPSS.sav [DataSet1] - IBM SPSS Statistics Data Editor

File Edit View Data Transform Analyze Graphs Utilities Add-ons

	Name	Label	Values
1	Person		None
2	Gender		{1, Male}...
3	Heardofcon...	Has the respondent heard of the condition	{0, N/A}...
4	Set	Order	None
5	@1st	Favourite video	{1, Gene...
6	@2nd	2nd favourite	{1, Vide...
7	@3rd	3rd favourite	{1, Vide...
8	@4th	Least favourite	{1, Vide...
9	Combination	Order videos seen	None
10	VideoAGen...	General video A understanding	{1, Stron...
11	VideoBdoct...	Doctors video B understanding	{1, Stron...
12	VideoCOldU...	Old video C understanding	{1, Stron...
13	DEMOUnde...	Demonstration D understanding	{1, Stron...
14	TotalAGen	Overall score (video A)	None
15	TotalBdoc	Overall score (video B)	None
16	TotalCOld	Overall score (video C)	None
17	TotalDDEMO	Overall score (demo D)	None
18			

Data View Variable View

FIGURE 7.3 *SPSS variable label.*

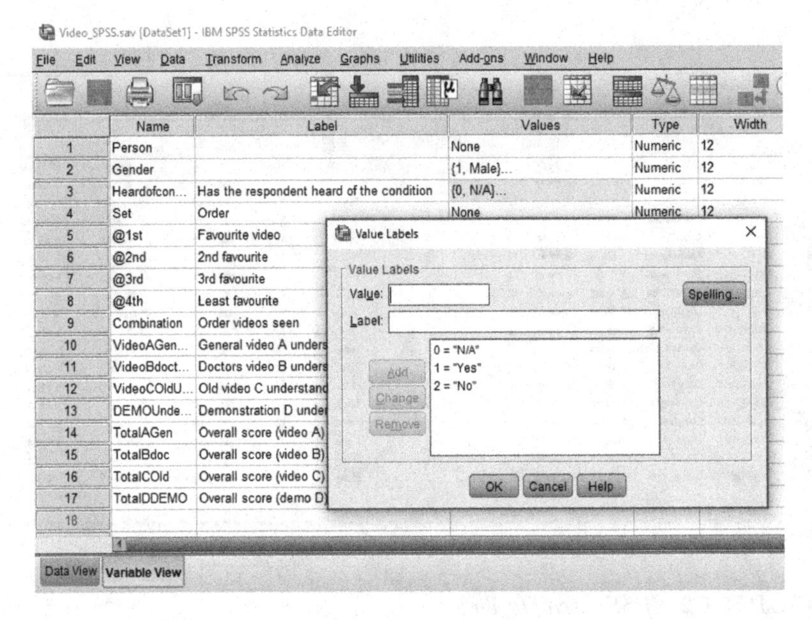

FIGURE 7.4 *SPSS value label.*

The R package {haven} allows us to capture all of this information.

7.1.2.1 Read SPSS data into R

For the first step, we will read in the data from the SPSS file with the default parameters.

```
df_video <- read_spss(dpjr::dpjr_data("Video_SPSS.sav"))

# display the data in the console
df_video
```

```
## # A tibble: 20 x 17
##    Person Gender     Heardofcondition   Set `@1st`
##     <dbl> <dbl+lbl>  <dbl+lbl>        <dbl> <dbl+lbl>
##  1      1 1 [Male]    1 [Yes]              1 4 [Demonst~
##  2      2 2 [Female]  1 [Yes]              1 4 [Demonst~
##  3      3 2 [Female]  2 [No]               1 4 [Demonst~
##  4      4 2 [Female] NA                    1 4 [Demonst~
##  5      5 2 [Female]  2 [No]               1 4 [Demonst~
##  6      6 2 [Female]  1 [Yes]             11 1 [General~
##  7      7 2 [Female]  1 [Yes]             11 1 [General~
```

```
## 8      8 1 [Male]     1 [Yes]            11 1 [General~
##  9      9 2 [Female] NA                 11 1 [General~
## 10     10 2 [Female] NA                 11 1 [General~
## 11     11 1 [Male]   NA                 24 3 [Old vid~
## 12     12 2 [Female] NA                 24 3 [Old vid~
## 13     13 1 [Male]   NA                 24 3 [Old vid~
## 14     14 2 [Female] NA                 24 3 [Old vid~
## 15     15 2 [Female] NA                 24 3 [Old vid~
## 16     16 1 [Male]     2 [No]           10 2 [Medical~
## 17     17 1 [Male]     2 [No]           10 2 [Medical~
## 18     18 2 [Female]   1 [Yes]          10 2 [Medical~
## 19     19 2 [Female]   1 [Yes]          10 2 [Medical~
## 20     20 1 [Male]     1 [Yes]          10 2 [Medical~
## # i 12 more variables: `@2nd` <dbl+lbl>,
## #   `@3rd` <dbl+lbl>, `@4th` <dbl+lbl>,
## #   Combination <dbl>,
## #   VideoAGenUnderstandingCONDITION <dbl+lbl>,
## #   VideoBdoctorUnderstandingCONDITION <dbl+lbl>,
## #   VideoCOldUnderstandingCONDITION <dbl+lbl>,
## #   DEMOUnderstandingCONDITION <dbl+lbl>, ...
```

Some of the variable names have leading `@` (at sign)—they will cause us some headaches later, so let's use the `clean_names()` function from the {janitor} package (Firke, 2021)[3] to clean them up.

```
df_video <- janitor::clean_names(df_video)
```

Now, we can use the `look_for()` function from {labelled} to view the contents of the dataframe:

```
look_for(df_video)
```

7.1.2.2 Handling missing values

Very often, SPSS files will have "user defined missing values." In the video survey file, they have been coded as "NA". But often, the analysis of survey results will have multiple types of "missing values":

- respondent left the question blank

[3]The reference page for {janitor} is here: http://sfirke.github.io/janitor/

- respondent didn't answer the question because of skip logic (in other words, the respondent didn't see the question)
- the analyst may have decided to code "Don't know" or "Not applicable" as "missing" when calculating the percentages of responses in the other categories.

Depending on the circumstance, you may want to count some of these. For example, if there a lot of "Don't know" and "Not applicable" responses, you may wish to analyze which one it is (they mean very different things!) If they are all coded as "NA", you have lost the ability to gain that insight.

In the code below, adding the `user_na = TRUE` argument to the `read_spss()` function maintains the original values.

```
df_video <- read_spss(dpjr::dpjr_data("Video_SPSS.sav"), user_na = TRUE)

df_video <- janitor::clean_names(df_video)

df_video
```

```
## # A tibble: 20 x 17
##    person gender       heardofcondition   set x1st
##     <dbl> <dbl+lbl>    <dbl+lbl>        <dbl> <dbl+lbl>
##  1      1 1 [Male]     1 [Yes]              1 4 [Demonst~
##  2      2 2 [Female] 1 [Yes]              1 4 [Demonst~
##  3      3 2 [Female] 2 [No]               1 4 [Demonst~
##  4      4 2 [Female] 0 (NA) [N/A]         1 4 [Demonst~
##  5      5 2 [Female] 2 [No]               1 4 [Demonst~
##  6      6 2 [Female] 1 [Yes]             11 1 [General~
##  7      7 2 [Female] 1 [Yes]             11 1 [General~
##  8      8 1 [Male]     1 [Yes]             11 1 [General~
##  9      9 2 [Female] 0 (NA) [N/A]        11 1 [General~
## 10     10 2 [Female] 0 (NA) [N/A]        11 1 [General~
## 11     11 1 [Male]     0 (NA) [N/A]        24 3 [Old vid~
## 12     12 2 [Female] 0 (NA) [N/A]        24 3 [Old vid~
## 13     13 1 [Male]     0 (NA) [N/A]        24 3 [Old vid~
## 14     14 2 [Female] 0 (NA) [N/A]        24 3 [Old vid~
## 15     15 2 [Female] 0 (NA) [N/A]        24 3 [Old vid~
## 16     16 1 [Male]     2 [No]              10 2 [Medical~
## 17     17 1 [Male]     2 [No]              10 2 [Medical~
## 18     18 2 [Female] 1 [Yes]             10 2 [Medical~
## 19     19 2 [Female] 1 [Yes]             10 2 [Medical~
## 20     20 1 [Male]     1 [Yes]             10 2 [Medical~
## # i 12 more variables: x2nd <dbl+lbl>, x3rd <dbl+lbl>,
```

```
## #   x4th <dbl+lbl>, combination <dbl>,
## #   video_a_gen_understanding_condition <dbl+lbl>,
## #   video_bdoctor_understanding_condition <dbl+lbl>,
## #   video_c_old_understanding_condition <dbl+lbl>,
## #   demo_understanding_condition <dbl+lbl>,
## #   total_a_gen <dbl>, total_bdoc <dbl>, ...
```

What were "NA" in the first version are now "0".

7.1.2.3 Exploring the data

The `attributes()` function gives us a way to view the details of the variables.

First, we can look at the attributes of the whole dataframe.

```
attributes(df_video)
```

```
## $class
## [1] "tbl_df"     "tbl"        "data.frame"
##
## $row.names
##  [1]  1  2  3  4  5  6  7  8  9 10 11 12 13 14 15 16 17
## [18] 18 19 20
##
## $names
##  [1] "person"
##  [2] "gender"
##  [3] "heardofcondition"
##  [4] "set"
##  [5] "x1st"
##  [6] "x2nd"
##  [7] "x3rd"
##  [8] "x4th"
##  [9] "combination"
## [10] "video_a_gen_understanding_condition"
## [11] "video_bdoctor_understanding_condition"
## [12] "video_c_old_understanding_condition"
## [13] "demo_understanding_condition"
## [14] "total_a_gen"
## [15] "total_bdoc"
## [16] "total_c_old"
## [17] "total_ddemo"
```

Or we can use the dollar sign method of specifying a single variable; in this case, `heardofcondition`:

```
attributes(df_video$heardofcondition)
```

```
## $label
## [1] "Has the respondent heard of the condition"
##
## $na_values
## [1] 0
##
## $class
## [1] "haven_labelled_spss" "haven_labelled"
## [3] "vctrs_vctr"          "double"
##
## $format.spss
## [1] "F12.0"
##
## $display_width
## [1] 6
##
## $labels
## N/A Yes  No
##   0   1   2
```

In the output above, the `$label` is the question, and the `$labels` are the labels associated with each value, including N/A. Note that there's also a `$na_values` shown.

You can also string these together in our code. For example, if we want to see the value labels and nothing else, the code would be as follows:

```
attributes(df_video$heardofcondition)$labels
```

```
## N/A Yes  No
##   0   1   2
```

7.1.3 Factors

Factors behave differently than labels: they don't preserve both the label and the value. Instead they display the value and preserve the sort order (that can be either automatically set or defined in the code).

In the code below, we will set all of the variables in the `df_video` dataframe as factors. In the first code chunk there are no additional parameters.

```
(df_video_factor <- as_factor(df_video))
```

```
## # A tibble: 20 x 17
##    person gender heardofcondition   set x1st       x2nd
##     <dbl> <fct>  <fct>            <dbl> <fct>      <fct>
##  1      1 Male   Yes                  1 Demonstr~ Vide~
##  2      2 Female Yes                  1 Demonstr~ Vide~
##  3      3 Female No                   1 Demonstr~ Vide~
##  4      4 Female N/A                  1 Demonstr~ Vide~
##  5      5 Female No                   1 Demonstr~ Vide~
##  6      6 Female Yes                 11 General ~ Vide~
##  7      7 Female Yes                 11 General ~ Vide~
##  8      8 Male   Yes                 11 General ~ Vide~
##  9      9 Female N/A                 11 General ~ Vide~
## 10     10 Female N/A                 11 General ~ Vide~
## 11     11 Male   N/A                 24 Old vide~ Vide~
## 12     12 Female N/A                 24 Old vide~ Vide~
## 13     13 Male   N/A                 24 Old vide~ Vide~
## 14     14 Female N/A                 24 Old vide~ Vide~
## 15     15 Female N/A                 24 Old vide~ Vide~
## 16     16 Male   No                  10 Medical ~ Demo~
## 17     17 Male   No                  10 Medical ~ Demo~
## 18     18 Female Yes                 10 Medical ~ Demo~
## 19     19 Female Yes                 10 Medical ~ Demo~
## 20     20 Male   Yes                 10 Medical ~ Demo~
## # i 11 more variables: x3rd <fct>, x4th <fct>,
## #   combination <dbl>,
## #   video_a_gen_understanding_condition <fct>,
## #   video_bdoctor_understanding_condition <fct>,
## #   video_c_old_understanding_condition <fct>,
## #   demo_understanding_condition <fct>,
## #   total_a_gen <dbl>, total_bdoc <dbl>, ...
```

In this second chunk, the `levels` parameter is set to `both`.

```
(df_video_both <- as_factor(df_video, levels="both"))
```

```
## # A tibble: 20 x 17
##    person gender     heardofcondition   set x1st  x2nd
```

```
##      <dbl> <fct>         <fct>                     <dbl> <fct> <fct>
##  1       1 [1] Male      [1] Yes                       1 [4] ~ [1] ~
##  2       2 [2] Female [1] Yes                          1 [4] ~ [1] ~
##  3       3 [2] Female [2] No                           1 [4] ~ [1] ~
##  4       4 [2] Female [0] N/A                          1 [4] ~ [1] ~
##  5       5 [2] Female [2] No                           1 [4] ~ [1] ~
##  6       6 [2] Female [1] Yes                         11 [1] ~ [3] ~
##  7       7 [2] Female [1] Yes                         11 [1] ~ [3] ~
##  8       8 [1] Male      [1] Yes                      11 [1] ~ [3] ~
##  9       9 [2] Female [0] N/A                         11 [1] ~ [3] ~
## 10      10 [2] Female [0] N/A                         11 [1] ~ [3] ~
## 11      11 [1] Male      [0] N/A                      24 [3] ~ [1] ~
## 12      12 [2] Female [0] N/A                         24 [3] ~ [1] ~
## 13      13 [1] Male      [0] N/A                      24 [3] ~ [1] ~
## 14      14 [2] Female [0] N/A                         24 [3] ~ [1] ~
## 15      15 [2] Female [0] N/A                         24 [3] ~ [1] ~
## 16      16 [1] Male      [2] No                       10 [2] ~ [4] ~
## 17      17 [1] Male      [2] No                       10 [2] ~ [4] ~
## 18      18 [2] Female [1] Yes                         10 [2] ~ [4] ~
## 19      19 [2] Female [1] Yes                         10 [2] ~ [4] ~
## 20      20 [1] Male      [1] Yes                      10 [2] ~ [4] ~
## # i 11 more variables: x3rd <fct>, x4th <fct>,
## #   combination <dbl>,
## #   video_a_gen_understanding_condition <fct>,
## #   video_bdoctor_understanding_condition <fct>,
## #   video_c_old_understanding_condition <fct>,
## #   demo_understanding_condition <fct>,
## #   total_a_gen <dbl>, total_bdoc <dbl>, ...
```

7.2 Creating a labelled dataframe from an Excel file

You might encounter a circumstance where the data collectors have originally used SPSS, SAS, or Stata{Stata}, but share the data as an Excel file. This might be because the receiving organization does not have the proprietary software, and the ubiquity of Excel as an analytic tool makes it a functional choice.

Typically in these circumstances, the data collectors make the data available in two parts:

- The individual records, with the variables coded numerically.

- A separate "code book", with the labels associated with each numeric value.

In the following example, the Palmer penguins data (Horst, 2020) has been stored in an Excel file. The first sheet in this Excel file contains the data, where the values in the three character variables (species, island, and sex) have been converted to numeric codes. For example, for the species variable, the Adelie penguins are represented by the value "1", Chinstrap are represented by "2", and Gentoo are represented by "3".

```
penguins_path <- dpjr::dpjr_data("penguins_labelled.xlsx")

penguins_data <- read_excel(penguins_path, sheet = "penguins_values")

head(penguins_data)
```

A glance at the dataframe shows that the variables `species`, `island`, and `sex` are all coded as numeric variables. Unlike the ".sav" file we read earlier in this chapter, this dataframe does not carry the labels as part of the variable.

We could write some code to apply labels and change the variable type, using the `val_labels()` function from the {labelled} package:

```
# assign the species names as labels
labelled::val_labels(penguins_data$species) <-
  c("Adelie" = 1, "Chinstrap" = 2, "Gentoo" = 3)

# view the labels
val_labels(penguins_data$species)

# another way to view the labels
attr(penguins_data$species, "labels")
```

We can also use R to read in the values and apply them programmatically. This will save time and effort and reduce the risk of typographical errors, particularly if there are many variables and those variables have many values to label.

If we return to the Excel file "penguins_labelled.xlsx" we find that the second sheet in the Excel file is called "penguins_codebook_source". This sheet contains the output SPSS creates when the "DISPLAY DICTIONARY" syntax (or the "File > Display Data File Information > Working File" GUI menu sequence) in SPSS is used to produce the "dictionary" (or codebook). There is information about each of the variables, including whether that variable is nominal or scale.

Of particular interest to us are the values associated with each variable. This information is saved under the heading "Variable Values"; for the penguins data file, this appears in row 36 of the Excel sheet.

Variable Values

Value		Label
species	1	Adelie
	2	Chinstrap
	3	Gentoo
island	1	Biscoe
	2	Dream
	3	Torgersen
sex	1	female
	2	male

FIGURE 7.5 *"Variable Values" from penguins SPSS output.*

First, we read the contents of a specific rectangle of data in the second sheet in the Excel file, containing the code book. Note that we start reading at row 37, so that "Value" and "Label" are assigned as our variable names. In this code, the "anchored" approach is used, where the upper left corner Excel cell is specified, and then the number of rows and columns to be read in the `dim` = argument.

```
penguins_path <- dpjr::dpjr_data("penguins_labelled.xlsx")

penguins_code <- read_excel(penguins_path,
                            sheet = "penguins_codebook",
                            range = anchored("A37", dim = c(9, 3)))

penguins_code
```

We have read the table as output by SPSS and saved in the Excel file, but there is additional wrangling required.

The first step in the pipe is to rename the variables. Not only has the name `Value` been given to the "Variable" column, but the name of the value column has been interpreted as blank and has been assigned the variable name `...2`.

The second step is to use the {dplyr} `fill()` function to populate the missing values in the `variable` variable. The default direction of the fill is downwards, so we don't need to specify the direction as an argument. We will examine other options for cleaning data later.

```
penguins_code <- penguins_code |>
  # rename variables
  rename(Value = "variable",
    ...2 = "variable_value",
    Label = "variable_label") |>
  # fill in blank names
  fill(variable) |>
  # change to numeric
  mutate(variable_value = as.numeric(variable_value))

penguins_code
```

At this point, there are two possible solutions:

1. Join the values from the "penguins_code" dataframe, so that in addition to the variable with numeric values, there is also a variable with the character strings.
2. Using the functions in the {labelled} package, convert the variables into "labelled" type, so the the numeric values remain and the labels are applied. In this scenario, the end-state is the same as reading the SPSS file directly.

Final step: join variables

```
penguins_code_wide <- penguins_code |>
#  mutate(row_num = row_number()) |>
  pivot_wider(names_from = c(variable), values_from = variable_label)

penguins_code_wide
```

7.3 Creating a labelled dataframe from SPSS syntax

In the previous chapter, we imported data files that had been created using the software packages SAS, SPSS, and Stata. We also saw how to work with *labelled* variables.

In our quest for data to analyze, we might come across a circumstance where an organization has collected survey data, and the data collectors have published the individual records, allowing researchers like you and me to further explore the data.

These data files contain individual records, rather than a data file that has been summarized with counts and percentages. A file with the individual records is sometimes referred to as a "micro-file", and one that has been anonymized for publication might be described as a Public-Use Micro File (abbreviated as "PUMF") or Public Use Microdata Sample (PUMS)\index{PUMS|see {Public-Use Micro File}}.

In this chapter, we will look at two circumstances where the data collectors have done just that. But rather than releasing the micro-data in a variety of formats, they have published a bundle that contains the raw text flat file along with syntax (code) files that apply variable names, variable labels, and value labels. This is great if we have a license for one of those proprietary software tools[4], but what if we are an R user?

Fortunately for us, there is a package that has functions to read a raw text file, and apply SPSS syntax to create an R object with labelled variables: {memisc} (Elff et al., 2021).

```
library(memisc)
```

7.3.1 National Travel Survey (NTS)

Our first example uses information from the National Travel Survey (NTS), collected and published by Statistics Canada (Statistics Canada, 2021).

Step 1 - Download the "SPSS" zip file for the 2020 reference period, and unzip it in the project sub-folder "data". This should create a sub-folder, so that the file path is now "data\2020-spss"

The zip file does *not* include an SPSS-format data file (.sav). Instead, the folder has a fixed-width text file for each of the three survey components (person, trips, and visits), and corresponding SPSS syntax files that can be used to read the files into the correct variables and to apply the variable and value labels.

For this example, we will use the 2020 NTS "Person" file. The raw data file provided by Statistics Canada is "PERSON_NTS2020_PUMF.txt".

Step 2 - Use the SPSS syntax files (in R)

The bundle also includes separate SPSS syntax files with the ".sps" extension. Because Canada is a bilingual country, Statistics Canada releases two versions

[4]Or a friend we can bribe with the promise of a chocolate bar.

of the label syntax files, one in English and the other in French. They can be differentiated through the last letter of the file name, before the extension. These are:

- variable labels: "Person_NTS2020_Pumf_vare.sps" (The equivalent file with French variable labels is "Person_NTS2020_Pumf_varf.sps".)
- variable values: "Person_NTS2020_Pumf_vale.sps"
- missing values: "Person_NTS2020_Pumf_miss.sps"

This structure works very well with the {memisc} functions, and these files do not require any further manipulation. (As we will see in our second example, this is not always the case.)

Because Statistics Canada has made the data available from other years, we will consciously create flexible code that will permit us to rerun our code later with a minimum number of changes. Accordingly, we will assign the year of the data as an object.

```
nts_year <- "2020"
```

We can now use the object with the year number in a variety of ways, as we define the locations of the various input files. Note that this code uses the `glue()` function from the {glue} package to create the file name strings, and the `here()` function from the {here} package to determine the file path relative to our RStudio project location.

```
# define locations of input files

nts_year_format <- glue(nts_year, "-SPSS")

input_folder <- here("data", nts_year_format)
input_folder
```

```
## [1] "E:/github_book/data_preparation_journey/data/2020-SPSS"
```

```
layout_folder <- here("data", nts_year_format, "Layout_cards")
layout_folder
```

```
##       [1]       "E:/github_book/data_preparation_journey/data/2020-
SPSS/Layout_cards"
```

```
# use the above to create objects with file names and paths

data_file_1 <- glue("PERSON_NTS", nts_year, "_PUMF.txt")
data_file <- glue(input_folder, "/Data_Données/", data_file_1)
data_file
```

```
## E:/github_book/data_preparation_journey/data/2020-SPSS/Data_Données/PERSON_NTS2020_PUMF.txt
```

```
columns_file_1 <- glue("Person_NTS", nts_year, "_Pumf_i.sps")
columns_file <- glue(layout_folder, columns_file_1, .sep = "/")
columns_file
```

```
## E:/github_book/data_preparation_journey/data/2020-SPSS/Layout_cards/Person_NTS2020_Pumf_i.sps
```

```
variable_labels_1 <- glue("Person_NTS", nts_year, "_Pumf_vare.sps")
variable_labels <- glue(layout_folder, variable_labels_1, .sep = "/")

variable_values_1 <- glue("Person_NTS", nts_year, "_Pumf_vale.sps")
variable_values <- glue(layout_folder, variable_values_1, .sep = "/")

missing_values_1 <- glue("Person_NTS", nts_year, "_Pumf_miss.sps")
missing_values <- glue(layout_folder, missing_values_1, .sep = "/")
```

The next stage is to create a definition of the dataset using the `memisc::spss.fixed.file()` import procedure.

```
# read file, create datafile with nested lists
nts_person_2020 <- memisc::spss.fixed.file(
  data_file,
  columns_file,
  varlab.file = variable_labels,
  codes.file = variable_values,
  missval.file = missing_values,
  count.cases = TRUE,
  to.lower = TRUE
)
```

```
# convert the resulting object into a "data.set" format
nts_person_2020_ds <- memisc::as.data.set(nts_person_2020)
```

That dataset object is now converted into a tibble, using the `as_haven()` function. This tibble has all of the variable and value labels assigned, accessible via the functions in the {labelled} package.

```
nts_person_2020_hav <- memisc::as_haven(nts_person_2020_ds)
nts_person_2020_hav
```

```
## # A tibble: 79,771 x 13
##     pumfid verdate  refyear    quarter inf_q01 inf_q02
## *    <dbl> <chr>    <int+lbl>  <int+l> <int+l> <int+lb>
##  1 1006690 10/12/2~ 2020 [Yea~ 1 [Ref~ 1 [Yes] 35 [Ont~
##  2 1006691 10/12/2~ 2020 [Yea~ 1 [Ref~ 1 [Yes] 35 [Ont~
##  3 1006692 10/12/2~ 2020 [Yea~ 1 [Ref~ 1 [Yes] 24 [Que~
##  4 1006693 10/12/2~ 2020 [Yea~ 1 [Ref~ 1 [Yes] 35 [Ont~
##  5 1006694 10/12/2~ 2020 [Yea~ 1 [Ref~ 1 [Yes] 24 [Que~
##  6 1006695 10/12/2~ 2020 [Yea~ 1 [Ref~ 1 [Yes] 35 [Ont~
##  7 1006696 10/12/2~ 2020 [Yea~ 1 [Ref~ 1 [Yes] 13 [New~
##  8 1006697 10/12/2~ 2020 [Yea~ 1 [Ref~ 1 [Yes] 11 [Pri~
##  9 1006698 10/12/2~ 2020 [Yea~ 1 [Ref~ 1 [Yes] 35 [Ont~
## 10 1006699 10/12/2~ 2020 [Yea~ 1 [Ref~ 1 [Yes] 35 [Ont~
## # i 79,761 more rows
## # i 7 more variables: page_grp <int+lbl>,
## #   gend_bin <int+lbl>, incomgr2 <int+lbl>,
## #   memc18up <int+lbl>, memclt18 <int+lbl>,
## #   resprov <int+lbl>, pumfwp <dbl>
```

In our final step, we save the dataframe as SPSS, Stata, and RDS files for future use.

One important note for Stata users is that there is a "version" argument, which allows us to create a file that can be read by older versions of Stata (the default is Stata 15).

```
# export haven format table as SPSS .sav file
haven::write_sav(nts_person_2020_hav,
                 here("data_output", "nts_person_2020.sav"))

# export haven format table as Stata .dta file
```

```
haven::write_dta(nts_person_2020_hav,
                 here("data_output", "nts_person_2020.13.dta"),
                 version = 13)


# export RDS
readr::write_rds(nts_person_2020_hav,
                 here("data_output", "nts_person_2020.rds"))
```

8 Importing data: PDF files

In this chapter:

- Reading data and text from PDF files
- Manipulating the resulting objects into a tidy structure

8.1 PDF files

> Portable Document Format (PDF) is a file format developed by Adobe in 1992 to present documents, including text formatting and images, in a manner independent of application software, hardware, and operating systems[1].

The PDF file format provides a great deal of functionality and flexibility to create high quality output. PDF files are often used for things like branded marketing materials and annual reports, where there is a use-case for a document that looks on screen as it would if printed on paper, with a combination of text and images, all designed and formatted in a way to make it visually appealing.

Included in those documents may be data we want for our analysis. Or we may want to analyze the text in the document. The way a PDF file is structured can make extracting the contents, whether data tables or text, a challenge. The "export to Excel" function in Adobe Acrobat and other PDF tools might work successfully; always give this a try first. The operative phrase in that sentence

[1]Wikipedia entry, "PDF" (Wikipedia contributors, 2022b)

is *might work*. Similar to the challenges that we confront when importing a formatted Excel file, your PDF exporter will confront structures that look like tables visually, but which are not in a consistent tidy format. Sometimes text wrapping in variable names leads to splitting into multiple rows, and in other instances what looks like multiple cells in the table are read as one.

Fortunately for us, other data scientists have already confronted the challenge of reading data tables in PDF files, and have made tools available for the rest of us. One such package is {pdftools}(Ooms, 2022)

In addition to {pdftools}, for this chapter we will also be using the data manipulation packages {dplyr} and {tidyr}, and the text manipulation package {stringr}(Wickham, 2019b)

```
# reading the contents of a PDF file
library(pdftools)

# data wrangling packages
library(dplyr)
library(tidyr)

# string manipulation
library(stringr)
```

8.1.1 Getting started

To get started in reading data from a PDF file, we will read data from a summary of the {palmerpenguins} (Horst et al., 2022; Horst, 2020) data.

While this is for a single table on a single page, the same approach will work for multiple tables across multiple pages, if the structure of each table on each page is the same. The process is always the same: start with the smallest unit and expand out.

- The way that PDF files are read, the smallest unit is a single row.
- Expand that to the entire table. Apply column splitting and assigning headers at the end of this stage.
- If there are multiple tables with the same structure, tackle that next by looping through each of the tables.
- And once you have a single page working, you can loop through multiple pages.

We read the entire contents of the file using the `pdf_text()` function from the {pdftools} package, and assign the contents to an R object in our environment.

Next we use the `str()` function to display a summary of the structure of our new object.

```
penguin_summary <- pdf_text(dpjr::dpjr_data("penguin_summary.pdf"))
```

```
str(penguin_summary)
```

```
 chr "  species   island      mean_flipper_length_mm mean_body_mass_g\n\n1 Adelie    Biscoe
188.795"| __truncated__
```

FIGURE 8.1 *penguin data from PDF file.*

What we have as a result of this is a single character string, with the contents of the file. In this example, there is only one page and thus one string. If you are dealing with a multi-page document, each page gets its own string.

In a visual scan of this output, we see the variable names starting with `species`. After the last of the variables `mean_body_mass_g` there are two end of line characters, represented by "\n" (pronounced "backslash n"). We will be using these end of line characters as part of the parsing process.

Another way to look at the contents of a page is to use the `cat()` function (from {base R}), which displays the contents as it would be printed:

```
cat(penguin_summary)
```

```
##   species   island      mean_flipper_length_mm mean_body_mass_g
##
## 1 Adelie    Biscoe                    188.7955         3709.659
##
## 2 Adelie    Dream                     189.7321         3688.393
##
## 3 Adelie    Torgersen                 191.1961         3706.373
##
## 4 Chinstrap Dream                     195.8235         3733.088
##
## 5 Gentoo    Biscoe                    217.1870         5076.016
```

Next we use the strsplit() function (from {base R}) to separate the long string into separate rows. As we noted there are two "\n" character pairs marking the end of each line; if we use only one in the strsplit() function, an extra row appears between the data rows.

```
penguin_summary_2 <- penguin_summary |>
  strsplit(split = "\n\n")

penguin_summary_2
```

```
## [[1]]
## [1] "  species   island      mean_flipper_length_mm mean_body_mass_g"
## [2] "1 Adelie    Biscoe                   188.7955         3709.659"
## [3] "2 Adelie    Dream                    189.7321         3688.393"
## [4] "3 Adelie    Torgersen                191.1961         3706.373"
## [5] "4 Chinstrap Dream                    195.8235         3733.088"
## [6] "5 Gentoo    Biscoe                 217.1870         5076.016\n"
```

At this stage, we need to apply some cleaning. If we don't do it now, it's going to cause some challenges later. There is an extra line return at the end of the last data cell, so let's use a regular expression to remove it. There's also space at the beginning of the first line and the number-space combination at the beginning of the data rows.

And by using the double square bracket accessor in the first row, the object returned is no longer nested, but instead has a single character string for each row.

```
penguin_summary_3 <- penguin_summary_2[[1]] |>
    str_remove("\\n$") |>
  str_remove("^\\s+") |>
  str_remove("^\\d ")
```

Next, we create a list object with the variable names, in order to keep the names separate from the data in our resulting table. This is accomplished by assigning the row that contains the names (row 1, identified using the square bracket [1]) to a vector with one element.

```
penguin_variable_vector <- as.vector(penguin_summary_3[1])

penguin_variable_vector
```

```
## [1] "species   island      mean_flipper_length_mm mean_body_mass_g"
```

The next step is to split this vector into 4 separate elements at the points where there are one or more white spaces, identified in the regular expression "\s+". The "\s" finds a single space, and adding the "+" extends that to "one or more" of the character the precedes it.

This returns another nested list.

```
penguin_variable_list <- str_split(penguin_variable_vector, "\\s+")
penguin_variable_list
```

```
## [[1]]
## [1] "species"                "island"
## [3] "mean_flipper_length_mm" "mean_body_mass_g"
```

In order to have the structure we need, we apply the `unlist()` function to the result.

```
penguin_variable_names <- unlist(penguin_variable_list)
penguin_variable_names
```

```
## [1] "species"                "island"
## [3] "mean_flipper_length_mm" "mean_body_mass_g"
```

It is possible to turn those two steps into one by wrapping the string split inside the `unlist()`:

```
penguin_variable_names <-
  unlist(str_split(penguin_variable_vector, "\\s+"))
penguin_variable_names
```

```
## [1] "species"                "island"
## [3] "mean_flipper_length_mm" "mean_body_mass_g"
```

The next step is to build a table, using a for-loop. (See (Wickham et al., 2023b, Chapter 27 Iteration) for a review of this process.)

```
# 1. define output
penguin_table <- tibble(value = NULL)

# 2. loop through the rows
# Because the first element is the titles,
# we want the length to be one shorter
loop_length <- length(penguin_summary_3) - 1

for (i in 1:5) {
  j <- i + 1
  dat <- as_tibble(penguin_summary_3[j])
  penguin_table <- bind_rows(penguin_table, dat)
}

penguin_table
```

```
## # A tibble: 5 x 1
##   value
##   <chr>
## 1 Adelie    Biscoe                     188.7955             3709.659
## 2 Adelie    Dream                      189.7321             3688.393
## 3 Adelie    Torgersen                  191.1961             3706.373
## 4 Chinstrap Dream                      195.8235             3733.088
## 5 Gentoo    Biscoe                     217.1870             5076.016
```

The loop has created a tibble with one column, `value`. We will apply the `separate()` function to split the value into each of the variable names we assigned to the object `penguin_variable_names` earlier. The split is based on the location of one or more spaces in the original string, using the "\s+" regular expression.

```
penguin_table <- tidyr::separate(
  penguin_table,
  value,
  into = penguin_variable_names,
  "\\s+"
  )

penguin_table
```

```
## # A tibble: 5 x 4
##   species   island    mean_flipper_length_mm mean_body_mass_g
```

```
##   <chr>     <chr>     <chr>                  <chr>
## 1 Adelie    Biscoe    188.7955               3709.659
## 2 Adelie    Dream     189.7321               3688.393
## 3 Adelie    Torgersen 191.1961               3706.373
## 4 Chinstrap Dream     195.8235               3733.088
## 5 Gentoo    Biscoe    217.1870               5076.016
```

There is one final step: to change the variable types of the values to numeric.

```
penguin_table <- penguin_table |>
  mutate(
    across(
    contains("mean_"),
    ~ as.numeric(.x)
    )
  )

penguin_table
```

```
## # A tibble: 5 x 4
##   species   island    mean_flipper_length_mm mean_body_mass_g
##   <chr>     <chr>                      <dbl>            <dbl>
## 1 Adelie    Biscoe                      189.            3710.
## 2 Adelie    Dream                       190.            3688.
## 3 Adelie    Torgersen                   191.            3706.
## 4 Chinstrap Dream                       196.            3733.
## 5 Gentoo    Biscoe                      217.            5076.
```

8.1.2 Extended example: ferry traffic

For this exercise, we will extract a data table from *Annual Report to the British Columbia Ferries Commissioner* for the fiscal year ending March 31, 2021 (that is, the twelve month period from 2020-04-01 to 2021-03-31) (British Columbia Ferry Services Inc., 2021).

This report is published as a PDF file that is 101 pages long and contains a mixture of written text and data tables.

{pdftools} gives us the ability to download and read the file from a local copy or from the web. The chunk below shows the code to download the file from a URL.

```
# download file from web URL
download.file(
  url = "https://www.bcferries.com/web_image/h29/h7a/8854124527646.pdf",
  destfile = "bcferries_2021.pdf",
  mode = "wb"
  )
```

Once the file is downloaded, we can then read the file using the `pdf_text()` function and assign the contents to an R object in our environment. Of course, we also have the option to read a file that we've already downloaded.

```
# alternative, from the {dpjr} package
bcf <- pdf_text(dpjr::dpjr_data("bcferries_2021.pdf"))
```

The code above produces a vector with 101 character strings—one for each page of the original document. We can access each page, using the square bracket accessor syntax of R. Here is the code to see the contents of the first page:

```
bcf[1]
```

```
[1] "British Columbia Ferry Services Inc.\n\n            Annual Report\n              to the\nBritish
Columbia Ferries Commissioner\n\n    Year Ended March 31, 2021\n"
```

FIGURE 8.2 *BC Ferries title page from PDF file.*

Using the `cat()` function:

```
cat(bcf[1])
```

```
## British Columbia Ferry Services Inc.
##
##               Annual Report
##                   to the
## British Columbia Ferries Commissioner
##
##     Year Ended March 31, 2021
```

If we compare these two outputs, we see that the first shows the characters "\n", which indicates the end of a line. These have been used to render the layout in the `cat()` version.

The period covered by this annual report captures the profound impact on travel during the first year of the COVID-19 pandemic, and the report has comparisons in vehicle and passenger volumes with the previous, pre-pandemic year. Let's pull the data from the table that compares the number of passengers by each of the routes in the system. This table is on page 11 of the PDF file:

Operations Summary Report for the Year Ended March 31, 2021

	K	L	M	N	O	P	% Sailings Within 10 Min. (Note 3)		
Routes	Passengers Fiscal 2021	Passengers Fiscal 2020	Passenger Growth (K - L)	Passenger Tariff Revenue Fiscal 2021 Note 2	Passenger Tariff Revenue Fiscal 2020 Note 2	Passenger Tariff Revenue Growth (N - O)	YE Fiscal 2019	YE Fiscal 2020	YE Fiscal 2021
1	2,578,221	6,124,234	(3,546,013)	39,230,411	91,309,787	(52,079,376)	87.6%	87.2%	86.6%
2	1,669,054	3,298,151	(1,629,097)	25,150,326	48,775,753	(23,625,427)	81.8%	87.3%	86.2%
3	1,797,894	2,631,102	(833,208)	9,517,331	13,518,977	(4,001,646)	81.4%	88.0%	83.7%
30	1,158,810	1,652,801	(493,991)	17,378,372	23,970,199	(6,591,827)	85.3%	83.2%	80.6%
Major Routes	**7,203,979**	**13,706,288**	**(6,502,309)**	**91,276,440**	**177,574,716**	**(86,298,276)**	**84.3%**	**86.6%**	**84.3%**
10	15,788	45,748	(29,960)	1,577,178	5,096,255	(3,519,077)	85.7%	91.8%	92.4%
11	17,385	47,070	(29,685)	543,176	1,522,607	(979,431)	92.4%	92.1%	93.8%
28	917	7,585	(6,668)	45,216	939,599	(894,383)	71.0%	72.1%	60.2%
Northern Routes	**34,090**	**100,403**	**(66,313)**	**2,165,570**	**7,558,461**	**(5,392,891)**	**85.0%**	**85.6%**	**82.8%**
4	462,546	662,431	(199,885)	1,924,700	2,742,617	(817,917)	93.9%	92.4%	96.4%
5	396,767	501,219	(104,452)	1,525,387	1,909,393	(384,006)	83.0%	81.4%	82.3%
6	395,352	479,800	(84,448)	1,381,288	1,644,946	(263,658)	74.1%	93.8%	96.1%
7	277,667	361,533	(83,866)	1,462,453	1,844,460	(382,007)	96.6%	95.8%	93.8%
8	924,639	1,281,422	(356,783)	2,731,011	3,687,853	(956,842)	91.7%	92.1%	95.5%
9	350,010	565,051	(215,041)	3,701,369	5,893,941	(2,192,572)	81.3%	77.9%	77.0%
12	108,826	198,874	(90,048)	441,337	812,625	(371,288)	86.1%	92.0%	95.0%
13	37,660	43,905	(6,245)	148,890	151,606	(2,716)	99.4%	99.7%	98.5%
17	258,792	396,958	(138,166)	2,719,390	4,020,165	(1,300,775)	92.1%	91.2%	94.4%
18	156,173	171,914	(15,741)	419,588	468,168	(48,580)	95.6%	94.0%	90.8%
19	599,439	805,907	(206,468)	1,598,175	2,015,334	(417,159)	85.9%	86.5%	77.7%
20	192,208	266,724	(74,516)	444,859	549,780	(104,921)	71.7%	68.9%	64.2%
21	460,556	542,837	(82,281)	1,128,583	1,274,810	(146,227)	97.8%	98.3%	98.4%
22	212,820	253,092	(40,272)	578,717	636,175	(57,458)	96.2%	97.9%	97.5%
23	690,845	871,056	(180,211)	1,642,909	2,055,963	(413,054)	97.0%	98.0%	97.3%
24	102,820	119,691	(16,871)	309,262	373,924	(64,662)	90.2%	89.2%	83.9%
25	170,795	243,899	(73,104)	540,173	751,754	(211,581)	86.5%	76.8%	82.1%
26	47,365	104,336	(56,971)	118,366	286,422	(168,056)	96.4%	95.9%	98.7%
Minor Routes	**5,845,280**	**7,870,649**	**(2,025,369)**	**22,816,457**	**31,119,936**	**(8,303,479)**	**89.3%**	**89.9%**	**89.8%**
Total	**13,083,349**	**21,677,340**	**(8,593,991)**	**116,258,467**	**216,253,113**	**(99,994,646)**	**88.5%**	**89.4%**	**89.1%**
Obligation deferred (settled)				-	-				
Total passenger revenue				**116,258,467**	**216,253,113**				
Total vehicle and passenger revenue				**424,077,117**	**613,201,638**				

FIGURE 8.3 *BC Ferries, passenger total.*

From the object we created from the PDF report, we can separate the contents of page 11 into its own object:

```
bcf_11 <- bcf[11]
```

```
bcf_11
```

Wait! This isn't the page we want! While the PDF file numbers the pages, the title page isn't numbered. The page numbered "11" is in fact the twelfth page in the file.

```
[1] "Operations Summary Report for the Year Ended March 31, 2021\n                                                    A
B                                   C                  D                  E                        F
G                 H                 I                   J\n\n\n
Capacity                                                Capacity          Capacity
AEQ Tariff        AEQ Tariff        AEQ Tariff\n                                   Actual Round
AEQ's Carried                                                   AEQ's Carried        AEQ Growth\n      Routes
Provided                                           Utilization Fiscal    Utilization
Revenue Fiscal    Revenue Fiscal    Revenue Growth\n                                      Trips
Fiscal 2021                                                     Fiscal 2020        (C - F)\n
(AEQ's)                                            2021       (C / B)    Fiscal 2020
2021 Note 2       2020 Note 2          (H - I)\n\n\n\n      1                              3,250.0
2,239,426                         1,557,909                     69.6%              84.1%              2,369,436
(811,527) $ 106,232,291        $   152,416,489       (46,184,198)\n        2
2,213.5       1,374,776                              838,402             61.0%              68.8%
1,334,158         (495,756)      55,498,711               81,940,368        (26,441,657)\n        3
2,834.0       1,755,940                              1,081,426           61.6%              65.9%
```

FIGURE 8.4 *BC Ferries page 11 from PDF file.*

```
bcf_11 <- bcf[12]
```

```
bcf_11
```

```
[1] "Operations Summary Report for the Year Ended March 31, 2021\n\n                                  K
L              M                 N                  O                  P\n
Passenger\n                                                                   Passenger        Passenger Tariff
Passenger Tariff                                       % Sailings Within 10 Min. (Note 3)\n
Passengers           Passengers                                                          Tariff Revenue\n     Routes
Growth          Revenue Fiscal Revenue Fiscal\n                                  Fiscal 2021          Fiscal 2020
Growth        (N -\n                                                                     (K - L)         2021 Note 2
2020 Note 2\n
O)                  YE Fiscal 2019   YE Fiscal 2020   YE Fiscal 2021\n\n\n         1
2,578,221           6,124,234      (3,546,013)         39,230,411          91,309,787      (52,079,376)
87.6%               87.2%              86.6%\n         2                          1,669,054          3,298,151
(1,629,097)          25,150,326        48,775,753      (23,625,427)                       81.8%              87.3%
86.2%\n        3                          1,797,894           2,631,102          (833,208)          9,517,331
13,518,977        (4,001,646)                       81.4%              88.0%            83.7%\n        30
1,158,810           1,652,801         (493,991)         17,378,372          23,970,199       (6,591,827)
85.3%               83.2%              80.6%\n Major Routes                    7,203,979          13,706,288
(6,502,309)          91,276,440       177,574,716      (86,298,276)                       84.3%              86.6%
84.3%\n\n       10                                     15,788              45,748           (29,960)           1,577,178
```

FIGURE 8.5 *BC Ferries page 12 from PDF file.*

Now we have an object that is a single text string. All of the data we want is there, but it is going to take some finesse to extract the columns we want, and in a format that we can use.

Using the "\n" line break character, we can split this single value into as many lines as there are on the page.

```
# break into lines
bcf_11_lines <- bcf_11[[1]] |>
  strsplit(split = "\n")

# how many lines in the page?
length(bcf_11_lines[[1]])
```

```
## [1] 55
```

Now we have a list of 1, with 55 individual elements in it.

The first thing that I notice is that the table starts a few rows down, and more problematically, the header row with the variable names is split due to text wrapping. I'm going to make the decision to enter the variable names manually and focus on extracting the numbers. The numbers start at row 13, which we can access as follows:

```
bcf_11_lines[[1]][13]
```

```
[1] "                1                         2,578,221              6,124,234       (3,546,013)          39,230,411
91,309,787      (52,079,376)                              87.6%                         87.2%                86.6%"
```

FIGURE 8.6 *BC Ferries: single line.*

We can specify a range of rows; let's set up the first two rows as our tester sample.

```
tester <- bcf_11_lines[[1]][13:14]
```

So how will we split this into separate pieces? Here's where the {stringr} package and our knowledge of regular expressions comes into use. We want to split the string where there are multiple spaces. If we split at every space, we'd get lots of columns. So we are separating (or splitting) where the spaces occur.

The regular expression to find any white space is "\s", and to find any number of them we need to add the plus sign "+". Remember that in R, we need to escape the backslash, so our regex gets expresses as "\s+".

```
test_result <- str_split(tester, "\\s+")
test_result
```

```
## [[1]]
##  [1] ""              "1"            "2,578,221"    "6,124,234"
##  [5] "(3,546,013)"   "39,230,411"   "91,309,787"   "(52,079,376)"
##  [9] "87.6%"         "87.2%"        "86.6%"
##
```

```
## [[2]]
##  [1] ""              "2"             "1,669,054"     "3,298,151"
##  [5] "(1,629,097)"  "25,150,326"    "48,775,753"    "(23,625,427)"
##  [9] "81.8%"         "87.3%"         "86.2%"
```

This remains a single list, now with two elements—one for each row.

Below, the code creates an object with a list of the variable names. (Sometimes it's more effective to tackle these cleaning problems head-on with a manual solution, rather than spend time on a one-time programmatic solution.)

```
bcf_col_names <- c(
  "blank",
  "route",
  "passengers_fy2021",
  "passengers_fy2020",
  "passenger_growth",
  "passenger_tariff_revenue_fy2021",
  "passenger_tariff_revenue_fy2020",
  "passenger_tariff_growth",
  "pct_sailings_10_mins_fy2019",
  "pct_sailings_10_mins_fy2020",
  "pct_sailings_10_mins_fy2021"
)
```

```
as_tibble(test_result, .name_repair = "unique")
```

```
## New names:
## * `` -> `...1`
## * `` -> `...2`

## # A tibble: 11 x 2
##    ...1          ...2
##    <chr>         <chr>
##  1 ""            ""
##  2 "1"           "2"
##  3 "2,578,221"   "1,669,054"
##  4 "6,124,234"   "3,298,151"
##  5 "(3,546,013)" "(1,629,097)"
##  6 "39,230,411"  "25,150,326"
##  7 "91,309,787"  "48,775,753"
```

```
## 8 "(52,079,376)" "(23,625,427)"
## 9 "87.6%"         "81.8%"
## 10 "87.2%"        "87.3%"
## 11 "86.6%"        "86.2%"
```

```
bcf_table <- tibble(value = NULL)

for (j in 1:2) {
  dat <- as_tibble(tester[j])
  bcf_table <- bind_rows(bcf_table, dat)
}

bcf_table
```

```
## # A tibble: 2 x 1
##   value
##   <chr>
## 1 "      1                    2,578,221        6,124,234     (3,546,0~
## 2 "      2                    1,669,054        3,298,151     (1,629,0~
```

```
bcf_table_2 <- tidyr::separate(bcf_table, value,
                                into = bcf_col_names,
                                "\\s+")

glimpse(bcf_table_2, width = 65)
```

```
## Rows: 2
## Columns: 11
## $ blank                          <chr> "", ""
## $ route                          <chr> "1", "2"
## $ passengers_fy2021              <chr> "2,578,221", "1,669,054"
## $ passengers_fy2020              <chr> "6,124,234", "3,298,151"
## $ passenger_growth               <chr> "(3,546,013)", "(1,629,~
## $ passenger_tariff_revenue_fy2021 <chr> "39,230,411", "25,150,3~
## $ passenger_tariff_revenue_fy2020 <chr> "91,309,787", "48,775,7~
## $ passenger_tariff_growth        <chr> "(52,079,376)", "(23,62~
## $ pct_sailings_10_mins_fy2019    <chr> "87.6%", "81.8%"
## $ pct_sailings_10_mins_fy2020    <chr> "87.2%", "87.3%"
## $ pct_sailings_10_mins_fy2021    <chr> "86.6%", "86.2%"
```

Hooray! (I should have left in all the failed attempts I made, before getting to this result—a lot of attempts that either failed outright, or ones where I managed a step or two before getting stuck.)

Except...all of the variables are character type, because of the commas that are used as thousand separators and the percent signs.

Let's go back to the `tester` object and see what we can do to pull them out the commas using the {stringr} function `str_remove_all()`. (Note that `str_remove()` removes the first instance of the specified string.)

For the purpose of our exercise, we won't worry about the numbers in parentheses, which are the difference between the two previous variables (and since they are negative, they are represented using the accounting format style of being in parentheses.) We won't deal with the percentages, but the percent signs could be removed in the same way as the commas.

```
# remove all commas
str_remove_all(tester, ",")
```

```
[1] "      1                  2578221       6124234    (3546013)       39230411       91309787
(52079376)         87.6%             87.2%           86.6%"
[2] "      2                  1669054       3298151    (1629097)       25150326       48775753
(23625427)         81.8%             87.3%           86.2%"
```

FIGURE 8.7 *BC Ferries: remove commas.*

Now let's put all of this together to capture the full table.

```
# read the relevant rows and remove the commas
bcf_data <- bcf_11_lines[[1]][13:42]
bcf_data <- str_remove_all(bcf_data, ",")

# set up the variable (column) names
bcf_col_names <- c(
  "blank",
  "route",
  "passengers_fy2021",
  "passengers_fy2020",
  "passenger_growth",
  "passenger_tariff_revenue_fy2021",
  "passenger_tariff_revenue_fy2020",
  "passenger_tariff_growth",
  "pct_sailings_10_mins_fy2019",
```

```
  "pct_sailings_10_mins_fy2020",
  "pct_sailings_10_mins_fy2021"
)


# set up the final table
bcf_table <- tibble(value = NULL)

# bind the rows together
for (j in 1:30) {
  dat <- as_tibble(bcf_data[j])
  bcf_table <- bind_rows(bcf_table, dat)
}

bcf_table
```

```
## # A tibble: 30 x 1
##    value
##    <chr>
##  1 "      1                        2578221          6124234     (3546013) ~
##  2 "      2                        1669054          3298151     (1629097) ~
##  3 "      3                        1797894          2631102      (833208) ~
##  4 "     30                        1158810          1652801      (493991) ~
##  5 " Major Routes                  7203979         13706288    (6502309)  ~
##  6 ""
##  7 "     10                             15788            45748      (29960~
##  8 "     11                             17385            47070      (29685~
##  9 "     28                               917             7585        (666~
## 10 "Northern Routes                     34090           100403      (66313)~
## # i 20 more rows
```

```
bcf_table_2 <- tidyr::separate(bcf_table, value,
                                into = bcf_col_names,
                                "\\s\\s+")
```

```
bcf_table_2
```

It's not quite perfect, and some things need to be removed:

A tibble: 30 x 11

blank <chr>	route <chr>	passengers_fy2021 <chr>	passengers_fy2020 <chr>	passenger_growth <chr>
	1	2578221	6124234	(3546013)
	2	1669054	3298151	(1629097)
	3	1797894	2631102	(833208)
	30	1158810	1652801	(493991)
Major Routes	7203979	13706288	(6502309)	91276440
	NA	NA	NA	NA
	10	15788	45748	(29960)
	11	17385	47070	(29685)
	28	917	7585	(6668)
Northern Routes	34090	100403	(66313)	2165570

1-10 of 30 rows | 1-5 of 11 columns Previous 1 2 3 Next

FIGURE 8.8 *BC Ferries: table 2.*

- there's the route type sub-total rows
- a blank row
- the difference columns

```
bcf_table_3 <- bcf_table_2 |>
# succinct way to remove a row which has a single "NA"
  na.omit() |>
  select(route,
         starts_with("passengers_"),
         starts_with("passenger_tariff_rev")) |>
  # change variables to numeric
  mutate(passengers_fy2021 = as.numeric(passengers_fy2021),
         passengers_fy2020 = as.numeric(passengers_fy2020),
         passenger_tariff_revenue_fy2021 =
           as.numeric(passenger_tariff_revenue_fy2021),
         passenger_tariff_revenue_fy2020 =
           as.numeric(passenger_tariff_revenue_fy2020))
```

```
bcf_table_3
```

If we need to create sub-totals by route type, it would be better to have a separate variable for that. Here's where we can use the `case_when()` function from {dplyr}.

```
bcf_table_3 <- bcf_table_3 |>
  mutate(route_type = case_when(
    route %in% c("1", "2", "3", "30") ~ "major",
```

A tibble: 25 x 5

route <chr>	passengers_fy2021 <dbl>	passengers_fy2020 <dbl>	passenger_tariff_revenue_fy2021 <dbl>
1	2578221	6124234	39230411
2	1669054	3298151	25150326
3	1797894	2631102	9517331
30	1158810	1652801	17378372
10	15788	45748	1577178
11	17385	47070	543176
28	917	7585	45216
4	462546	662431	1924700
5	396767	501219	1525387
6	395352	479800	1381288

1-10 of 25 rows | 1-4 of 5 columns Previous 1 2 3 Next

FIGURE 8.9 *BC Ferries: table 3.*

```
    route %in% c("10", "11", "28") ~ "northern",
    TRUE ~ "minor"
  ))
```

```
bcf_table_3
```

A tibble: 25 x 6

route <chr>	passengers_fy2021 <dbl>	passengers_fy2020 <dbl>	passenger_tariff_revenue_fy2021 <dbl>
1	2578221	6124234	39230411
2	1669054	3298151	25150326
3	1797894	2631102	9517331
30	1158810	1652801	17378372
10	15788	45748	1577178
11	17385	47070	543176
28	917	7585	45216
4	462546	662431	1924700
5	396767	501219	1525387
6	395352	479800	1381288

1-10 of 25 rows | 1-4 of 6 columns Previous 1 2 3 Next

FIGURE 8.10 *BC Ferries: table 3, final.*

Now we can use our data wrangling skills to calculate the percent change between the two years. Let's look at the change in passenger volume.

```
bcf_table_3 |>
  group_by(route_type) |>
  summarize(total_fy2021 = sum(passengers_fy2021),
            total_fy2020 = sum(passengers_fy2020)) |>
```

```
  mutate(pct_change =
           (total_fy2021 - total_fy2020) / total_fy2020 * 100)
```

```
## # A tibble: 3 x 4
##   route_type total_fy2021 total_fy2020 pct_change
##   <chr>             <dbl>        <dbl>      <dbl>
## 1 major           7203979     13706288      -47.4
## 2 minor           5845280      7870649      -25.7
## 3 northern          34090       100403      -66.0
```

Our analysis: The COVID-19 pandemic had an *enormous* impact on travel. The fiscal year 2021 covers the period April 1, 2020 through March 31, 2021, a period starting almost exactly as the strict limitations to travel were imposed. Even though the restrictions started to lift later in the 12-month period, the travel on the major routes was down by nearly half.

9

Data from web sources

In this chapter:

- Importing data from web repositories

9.1 Introduction

There is a lot of valuable data out there on the internet, just waiting for curious data analysts to find nuggets of insight in it. There are a variety of ways to access that data; in some cases, you can download a CSV or Excel file and then read that into R.

But the brilliant people that make up the R community have made our life a whole lot easier by creating packages that allow us to write R scripts that link directly to data we want. This is particularly valuable if you are accessing a data set that gets updated regularly (say, a weekly financial report or the results of a monthly survey).

The foundation of this connection to data sources on the internet are APIs (for Application Programming Interface). APIs allow two pieces of software to communicate. In the context of data access, a site will be set up with an API that defines how data can be accessed and how it will be provided.

In general, an API package will require a connection to the internet. In some cases, an API may require a user key be requested (a strategy the site may use to authenticate users or limit the number or size of requests).

There are two sorts of R packages that utilize APIs. The first are those that are dedicated to a particular source of data, such as a statistics agency. In addition to getting the data, these packages often have additional functions which do a lot of the initial manipulation of the data for us, getting it into a format that is useful for R.

The second type are beyond the scope of this book, and are those that allow us to connect to the raw data source. Often with these sources, we will need to do some wrangling of the data before we can use it. If you are interested in learning more about creating your own connections in R to websites that provide an API, the R package {httr2}(Wickham, 2023) might be a good place to start.

9.2 R packages for direct connection

In this section, we will explore examples that allows access to the data repositories hosted by the UK Office of National Statistics and Statistics Canada. There are many other tools available; some are listed at the end of this chapter. Neither of these sites requires (at least at the time of writing) an API key.

9.2.1 Office of National Statistics (UK)

Our first example will be {onsr}, which downloads data directly from the UK's Office of National Statistics (Vasilopoulos, 2022).

```
library(onsr)
```

For this, we would like to obtain time series data of retail sales.

The first thing we might want to do is explore what's available in our area of interest. The `ons_datasets()` function gives a full list of the data that is currently available.

```
datasets <- ons_datasets()

datasets |>
  select(id, description) |>
  filter(str_detect(id, "retail"))
```

From this we will select "Retail sales data for Great Britain in value and volume terms, seasonally and non-seasonally adjusted", which has the `id` value of "retail-sales-index".

The function `ons_ids()` provides a list of the `id` column that is obtained from the `ons_datasets` function.

Another option is the `ons_browse_qmi()` function, which opens the relevant ONS page in your web browser software.

```
ons_browse_qmi(id = "retail-sales-index")
```

Finally, we will use the relevant `id` value in an `ons_get()` function to retrieve the data.

```
df_retail_sales <- ons_get(id = "retail-sales-index")
```

There are 42,400 rows in this data (as of 2023-06-14), with multiple variables for each month. Your analysis will require further filtering and selecting, but with the {onsr} package providing the access to the data, you are well on your way.

9.2.2 Statistics Canada

Statistics Canada, a department of the Government of Canada, provides a socioeconomic time series database known as "CANSIM". The data posted to the database is most of the aggregate data collected by Statistics Canada on a regular basis, such as the Labour Force Survey (from which we get updates on the unemployment rate), the Consumer Price Index Survey (about inflation), and health-related information from pregnancy and births to life expectancy and deaths.

The package {cansim} (von Bergmann and Shkolnik, 2021) uses the API that Statistics Canada has created in order to facilitate direct download of hundreds of data tables and thousands of individual time series.

```
library(cansim)     # access Statistics Canada's CANSIM data repository
```

One of the many tables related to international travel that is published in CANSIM is "Table: 24-10-0050-01 Non-resident visitors entering Canada, by country of residence"[1].

Here's a screenshot of the table, showing the number of people arriving by their continent of residence.

[1]The webpage for the table, showing the most recently available data, is here: https://www150.statcan.gc.ca/t1/tbl1/en/tv.action?pid=2410005001

Geography	Canada (map)				
Country of residence[2]	March 2023	April 2023	May 2023	June 2023	July 2023
	Visitors				
Non-resident visitors entering Canada	1,382,309	1,674,026	2,377,181	3,332,607	4,000,340
United States of America residents entering Canada	1,094,111	1,300,810	1,831,912	2,632,723	3,132,681
Residents of countries other than the United States of America entering Canada	288,198	373,216	545,269	699,884	867,659
Americas, countries other than the United States of America	74,654	92,600	90,704	116,030	141,622
North America, countries other than the United States of America	42,446	56,789	49,022	56,651	69,313
Central America	3,226	5,116	4,851	6,787	6,769
Caribbean	11,203	15,742	14,952	20,830	28,585
South America	17,779	14,953	21,879	31,762	36,955
Americas, n.o.s.[3]	0	0	0	0	0

FIGURE 9.1 *CANSIM table 25-10-0050-01.*

This table might be useful if you were developing a marketing campaign and you wanted to understand where tourists are coming from—the biggest markets and the fastest growing. But as shown, it is at too high a level; this is Canada by continent. For our purposes, we need to see specific country, those entering Canada via British Columbia, and a longer time series.

We could use the filters and fiddle around making sure we got the right selections and then download the data, but if we have to do that next month it is going to be inefficient. We could also first click on the "download options" button and select "download entire table..." but again, this is not the most efficient way to do this, since there are a few manual steps we have to take and then open the file into R and filter to select the series we want.

This is where the {cansim} package comes in.

Step 1: download the data.

- the `get_cansim()` function takes the table number as the parameter, and connects to CANSIM and downloads the data into an R dataframe.

```
travellers <- get_cansim("24-10-0050-01")
```

We will come back to filtering and variable selection along with working with date fields later, but here's the code in R using the functions of the {dplyr} package to filter for the British Columbia cases during the month of June 2019:

```
travellers_bc <-
travellers |>
  filter(GEO == "British Columbia",
         REF_DATE == "2019-06")

glimpse(travellers_bc, width = 65)
```

There are 287 rows, one for each of the categories in the `Country of residence` variable. It's important to note that these are not just countries, but the variable also includes subtotals. What if we just want a monthly report on a single continent? One option would be for us to filter the "COUNTRY" variable; here we filter for visitors from Europe.

```
travellers_bc |>
  filter(`Country of residence` == "Europe")
```

One of the features of CANSIM and other similar tables is that the individual series are given a single identifier. In the case of CANSIM, these are called the "vector numbers". Scroll to the right on this table, and you'll see a variable called `VECTOR`. The vector identifier (sometimes called "the vector number" even though it's a character string!) for European travellers entering Canada via British Columbia is v1277198032.

The {cansim} package also gives us a function to select a single vector number, so we can be more efficient still. We can speed up the download time *and* skip the filtering steps to get to the one we want. (Note that this function requires us to enter a start date, and converts the reference date variable to the YYYY-MM-DD format.)

```
get_cansim_vector("v1277198032", start_time = "2015-01-01")
```

This table has the shortcoming of not including the "GEO" or "Country of residence" variables, so we need to ensure that we have adequately documented what has been accessed.

The {cansim} package includes other very useful functions if you find yourself exploring Statistics Canada data sets.

The function get_cansim_table_overview() provides an abstract about the data table.

```
get_cansim_table_overview("24-10-0050-01")
```

9.3 Other R packages for direct access to data

The following is an incomplete list (in alphabetical order by name) of other packages that have been developed to provide direct access to a specific web resource.

- {bcdata} (Teucher et al., 2023) has tools to search, query, and download tabular and 'geospatial' data from the British Columbia Data Catalogue.
- {fredr} (Boysel and Vaughan, 2023) connects to the "Federal Reserve Economic Data", or FRED, hosted by the Federal Reserve Bank of St. Louis. This data repository has (according to Wikipedia) over 800,000 economic time series from various sources.
- {opendatatoronto} (Gelfand and City of Toronto, 2022) provides a direct connection to the City of Toronto Open Data Portal.
- The {readabs} (Cowgill et al., 2023) package downloads and tidies data from the Australian Bureau of Statistics.
- {RBNZ} (Watson, 2020) downloads data from the Reserve Bank of New Zealand Website.
- {tidycensus} (Walker et al., 2023) has tools that connect to the United States Census Bureau and returns tidyverse-ready dataframes with the option of adding geographic details to allow for mapping. (See also (Walker, 2023).)

10

Linking to relational databases

In this chapter:

- Importing data directly from relational databases (SQL)
- Using relational data as part of the data preparation workflow

10.1 Relational data

Relational databases are commonly used for data storage. Unlike a "flat file" which contains all of the variables, the data is stored in multiple tables, which are linked via common variables (known as the "keys"). This facilitates efficient storage, as information does not need to be duplicated in multiple rows. We have seen an example of a relational table earlier, where we built a concordance table for the administrative geography of England and Wales in 6.

In that example, all of the data was loaded into our computer's memory. In other instances, the size of the database, across multiple tables, will be such that it exceeds the capacity of your computer. In those cases, a server-based database will be established, and as the data analyst, we will need to retrieve only the variables and records that we need for our work. A database query can also include the calculation of summary statistics, which shifts the computational load to the server.

While SQL is the most commonly used query language, there are R packages that allow us to

- Connect to a database,
- Write the code for our queries in R,
- Run the query, and
- Do our work in an R environment.

Writing your code in an R Markdown or Quarto document also gives you the flexibility to write the code in SQL, if that's your preferred approach.

This exercise replicates the joins described in the "Relational data" chapter of *R for Data Science* by Hadley Wickham & Garrett Grolemund (Wickham and Grolemund, 2016). Instead of using the R {nycflights13} package (Wickham and RStudio, 2021), we will use a SQLite version of the same database.

In this database, there are five separate tables. The table `flights` in the database contains all 336,776 flights that departed from New York City in 2013. The data comes from the US Bureau of Transportation Statistics and is documented in `?flights`

The other tables in the database are:

- `airlines` lets you look up the full carrier name from its abbreviated code,
- `airports` gives information about each airport, identified by the `faa` airport code,
- `planes` gives information about each plane, identified by its `tailnum`,
- `weather` gives the weather at each NYC airport for each hour.

The tables are *related* to `flights` by the fact that they have variables in common. These are known as the "key" variables.

This diagram shows the relationships:

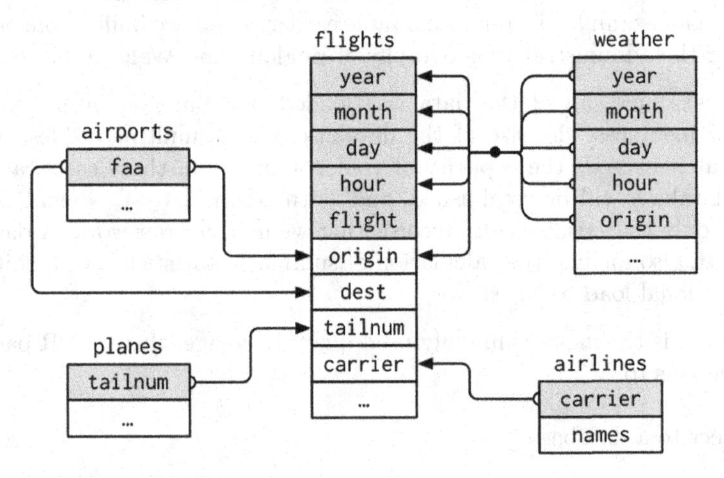

FIGURE 10.1 *nycflights13.*

(From (Wickham and Grolemund, 2016), p.174)

10.1.1 Connect to the database

SQL is a language widely used to manipulate and extract data in relational databases. As a consequence, there are many relational databases built in this format. Most often, these databases will be housed on a network server, but for smaller databases, you might install the file on your computer.

In R, we can use the package {dbplyr} (Wickham et al., 2023a) to access SQL databases and SQL functions. In addition, we need {DBI} (Müller, 2022) and {RSQLite} (Müller et al., 2023) to establish the connection to the RSQLite database.

```
library(dbplyr)
library(DBI)
library(RSQLite)
```

The code below establishes the connection to the database and assigns the connection (not the data table!) to the object `con_nycf` (for "Connection to New York City Flights"). You will note that the {RSQLite} function `SQLite()` is inside the {DBI} function `dbConnect()`. The {DBI} package supports a wide range of different database types, including the widely-used MySQL and Postgres.

```
# establish the connection to the database file
con_nycf <-
  DBI::dbConnect(RSQLite::SQLite(), "data/nycflights13_sql.sqlite")

# list the tables in the connected database
dbListTables(con_nycf)
```

```
## [1] "airlines"     "airports"     "flights"      "planes"
## [5] "sqlite_stat1" "sqlite_stat4" "weather"
```

Now that we have a connection to the database, we can establish a connection to a particular table, using the {dplyr} function `tbl()`. Note that the `flights` object is *not* the table but is the connection to the table.

```
flights <- tbl(con_nycf, "flights")
```

10.1.2 Submit queries

With the object "flights" now established in our environment, we can write R code to create a subset of the flights—those that went to Seattle. Again, the `flights_SEA` object is not a dataframe, but a set of instructions that creates the connection and the query.

```
flights_SEA <- flights |>
  filter(dest == "SEA")
```

We can also use the `show_query()` function of {dbplyr} to generate the SQL translation of the R code:

```
flights_SEA <- flights |>
  filter(dest == "SEA") |>
  show_query()
```

```
## <SQL>
## SELECT `flights`.*
## FROM `flights`
## WHERE (`dest` = 'SEA')
```

In SQL, we use `SELECT` to select the columns (or variables) we want (you will note this is the same term as {dplyr}). The asterisk "*" is a wildcard to select all the tables.

`FROM` indicates which table from which we want to draw the columns.

And finally, the filtering by city uses the SQL function `WHERE`.

In the code below, we create a summary table of the average flight time from New York to Seattle, by airline.

```
SEA <- flights_SEA |>
  select("month", "carrier", "air_time") |>
  group_by(carrier) |>
  summarize(
    n = n(),
    min_air_time = min(air_time),
    mean_air_time = mean(air_time),
    max_air_time = max(air_time)) |>
  # enter the resulting table into the R environment
  collect()
```

```
## Warning: Missing values are always removed in SQL aggregation functions.
## Use `na.rm = TRUE` to silence this warning
## This warning is displayed once every 8 hours.
```

```
SEA
```

```
## # A tibble: 5 x 5
##   carrier     n min_air_time mean_air_time max_air_time
##   <chr>   <int>        <dbl>         <dbl>        <dbl>
## 1 AA        365          289          336.          385
## 2 AS        714          277          326.          392
## 3 B6        514          283          330.          378
## 4 DL       1213          275          327.          389
## 5 UA       1117          280          326.          394
```

10.1.3 Using SQL in your R code

In addition to writing native R code, we can embed SQL inside R code. In the example below, an R chunk in the R Markdown document has R code that uses the `dbSendQuery()` function. Inside this function, we first name the connection we are using and then inside the quotation marks write SQL code: `"SELECT * FROM flights WHERE dest = 'SEA'"`.

This query instruction gets saved as the object `SEA_sql`.

The R chunk then has a second line that uses the `dbFetch()` function to run the SQL query.

Both the `dbSendQuery()` and `dbFetch()` are functions from the {DBI} package.

```
SEA_sql <- dbSendQuery(con_nycf,
                       "SELECT * FROM flights WHERE dest = 'SEA'")

dbFetch(SEA_sql)
```

10.2 Running SQL language chunks in R Markdown

The book *R Markdown: The Definitive Guide* (Xie et al., 2019, Chapter 2.7.3) provides instructions on how to set up your R Markdown in RStudio so that you can run SQL language chunks, including using SQLite.

Once this has been done, our R Markdown document can incorporate native SQL queries into the workflow. Note that in our SQL chunk, we specify the connection. As we continue to work through this example, this is the `con_nycf` object created earlier. The start of our SQL code chunk would contain this text:

```
{sql, connection=con_nycf}
```

```
SELECT *
FROM flights
WHERE dest = 'SEA'
```

10.2.1 Mutating joins and summary tables

A mutating join is one that combines variables from two tables, based on matching observations on *keys*.

In this R code, we create a summary table of the flights that went from New York to Seattle, by the name of the airline. The full airline name is not in the `flights` table; to get that, we need to join the airline name from the `airlines` table to the Seattle summary of the `flights` table.

The first step is to establish a connection to the `airlines` table.

```
# establish connection to airlines table
airlines <- tbl(con_nycf, "airlines")
```

Once the connection is made the tables can be joined using the `left_join()` function from {dplyr}, the grouped summary calculation made, and the table sorted from most to least frequent number of flights.

The table is also formatted for publication using the {gt} package (Iannone et al., 2023); the core `gt()` function is the first step, and many other formatting options are possible with this package.

```
# join and summary table
flights_SEA_summary <- flights_SEA |>
#  select(-origin, -dest) |>
  left_join(airlines, by = "carrier") |>
  group_by(name) |>
  tally() |>
  arrange(desc(n))

flights_SEA_summary |>
  gt()
```

name	n
Delta Air Lines Inc.	1213
United Air Lines Inc.	1117
Alaska Airlines Inc.	714
JetBlue Airways	514
American Airlines Inc.	365

The SQL code below returns the same table. We can see one difference in how the code is written—it runs "inside out" with the third step coming before the first and second, rather than in the linear manner we are accustomed to in our R pipes.

```
-- 3. select `name` variable from joined table and apply count
SELECT `name`, COUNT(*) AS `n`
FROM (
  SELECT `flights_sea`.*, `name`
-- 1. query to filter Seattle flights
  FROM (
    SELECT *
    FROM `flights`
    WHERE (`dest` = 'SEA')
  ) AS `flights_sea`
-- 2. join to airlines table
  LEFT JOIN `airlines`
    ON (`flights_sea`.`carrier` = `airlines`.`carrier`)
)
-- 4. define grouping for COUNT (from step 3) and sort
GROUP BY `name`
ORDER BY `n` DESC
```

TABLE 10.2 5 records

name	n
Delta Air Lines Inc.	1213
United Air Lines Inc.	1117
Alaska Airlines Inc.	714
JetBlue Airways	514
American Airlines Inc.	365

The joins are named using terms similar to those you are familiar with from {dplyr}. This left join will return all of the records from the `flights` table, and the variables from `airlines` where there is a match.

To indicate the key variable for the join, we use the SQL term `ON`. Note that we specify the table and the variable, separated by a period.

For more information on using SQL, with a focus on SQLite, Thomas Nield's *Getting Started with SQL* (Nield, 2016) is highly recommended.

10.3 Using the {tidylog} package

An important element of any table join is to check the result, to see if it conforms to our expectations. The {tidylog} package (Elders and Oldoni, 2020)

The authors of the package acknowledge that the functionality adds some computational and time overhead to the processing, but this may be worth the cost. Judiciously used, the information it returns can give an immediate indication if the code has worked. This is particularly true in the early stages of writing your code—the functions can be removed when you are confident you are getting the results you expect.

```
library("tidylog")
```

After running the `library(tidylog)` function, the results are automatically generated.

The Major League Baseball teams play in cities in the United States and Canada (currently only Toronto!), but draw the best players from around the world. In this example the code creates a summary table to show the country of birth of the player who batted in the 2001 season.

For this we will use the data stored in the R package that contains the Lahman baseball database (named after Sean Lahman, the person who initially created the database) (Friendly et al., 2020).

At the completion of this run, the {tidylog} package produces some summary information about the join, the grouping, and the tally functions.

```
## start with the Batting table
Lahman::Batting |>
  ## filter for year
  filter(yearID == "2001") |>
  ## join to People table (where the birth country is recorded)
  left_join(Lahman::People, by = "playerID") |>
  ## now group_by and tally
  group_by(birthCountry) |>
  tally() |>
  arrange(desc(n)) |>
  slice_head(n = 10)
```

```
## filter: removed 110,845 rows (99%), 1,339 rows remaining

## left_join: added 25 columns (birthYear, birthMonth, birthDay,
 birthCountry, birthState, …)

##            > rows only in x         0

##            > rows only in y  (19,456)

##            > matched rows       1,339

##            >                 ========

##            > rows total         1,339

## group_by: one grouping variable (birthCountry)

## tally: now 22 rows and 2 columns, ungrouped

## slice_head: removed 12 rows (55%), 10 rows remaining

## # A tibble: 10 x 2
##    birthCountry     n
##    <chr>        <int>
##  1 USA           1002
```

```
##  2 D.R.           120
##  3 P.R.            57
##  4 Venezuela       54
##  5 Mexico          19
##  6 Cuba            16
##  7 CAN             14
##  8 Japan           14
##  9 Panama          11
## 10 Australia        6
```

11

Exploration and validation strategies

In this chapter:

- Exploratory data analysis to identify problems, including the {skimr} package
- Using structured tests with the {validate} package

11.1 Identifying dirty data

Earlier in Chapter 2, challenges associated with "dirty data" were introduced.

The first challenge: How do we find the things that are problematic with our data?

The second challenge: What can and should we do about them?

The term "data validation" sometimes applies to verification that is applied when the data is collected (for example, in a web-based survey tool), and in other contexts it applies to a programmatic approach to ensuring the data meet specified conditions. In both cases, we can think of data validation as a form of quality control, from quick sanity checks to rigorous programmatic testing. (White et al., 2013)

Validation of data during collection might be through the use of drop-down boxes that eliminates spelling mistakes or entry fields that require the entered text to match a particular pattern. For example, a Canadian postal code is always in the pattern of capital letter (but not all letters), digit from 0 to 9, capital letter, space, digit, capital letter, and digit. The first letter of the postal codes specify a region; in Newfoundland and Labrador they start with the letter "A" while in Yukon they start with "Y", adding another dimension for possible validation. Electronic data collection can also enforce that an element is completed before submission.

These imposed validation rules come with risks and need to be thoughtfully implemented. Forcing a person to complete a field might lead to made-up values, which might be worse than a missing field. Imagine forcing someone to report their income—they might not be comfortable sharing that information, and in order to fulfill the requirements of the field enter a "1" or a preposterously very large number, neither of them accurate. In this case, it might be better for the form to provide a "Prefer not to answer" field, giving respondents an option that does not distort the final result.

For our work here, we will focus on checking and validating data after it has been collected. The analyst can specify parameters prior to it being made available for the next steps. These can be expressed as rules or assumptions, and are based on knowledge of the subject. (van der Loo and de Jonge, 2018, Chapter 6: Data Validation)

The type of things we might evaluate include:

- A field has to be structured in a certain way; examples include postal codes and date fields.
- A value might have a certain range or a limited number of categories.
- Some values, such as age, have to be positive.
- There might be a known relationship between two or more variables.

We now turn to approaches to evaluate the values in the data.

11.2 Exploratory data analysis

One way to identify dirty data is through the process of *exploratory data analysis* (EDA). (Tukey, 1977) The EDA process is, first and foremost, intended to help guide our analysis, but it has the additional benefit of providing clues about the ways that our data is dirty.

EDA consists of two groups of techniques:

- summary statistics
- data visualization

The things we are looking for at this stage include:

- the variable types

- missing values
- invalid values and outliers, including sentinel values
- data ranges that are too wide or too narrow
- the units of the data

(Adapted from (Zumel and Mount, 2019, p.54))

For this example, we will look at some synthetic data about the employees of a company. The data is stored in multiple sheets of an Excel file; we will explore the data in the sheet "df_HR_train", which records whether someone has completed the mandatory training courses or not.

```
readxl::excel_sheets(dpjr::dpjr_data("cr25/cr25_human_resources.xlsx"))
```

```
## [1] "readme"              "df_HR_main"          "df_HR_transaction"
## [4] "df_HR_train"
```

Once we have read the data into our environment, we can use some EDA methods to check the data.

```
df_HR_train <-
  readxl::read_excel(dpjr::dpjr_data("cr25/cr25_human_resources.xlsx"),
                     sheet = "df_HR_train")
```

One way we might begin our investigation is the `dplyr::glimpse()` function. This function returns the dimensions of the dataframe, the variable types, and the first few values of each variable.

```
glimpse(df_HR_train, width = 65)
```

```
## Rows: 999
## Columns: 5
## $ emp_id         <chr> "ID001", "ID002", "ID003", "ID004", "ID0~
## $ train_security <chr> "TRUE", "FALSE", "TRUE", "FALSE", "FALSE~
## $ train_crm      <chr> "TRUE", "FALSE", "TRUE", "TRUE", "FALSE"~
## $ train_qc       <chr> "NA", "FALSE", "FALSE", "FALSE", "FALSE"~
## $ train_history  <chr> "TRUE", "TRUE", "TRUE", "TRUE", "TRUE", ~
```

The variables that start with `train_` have been stored as character type `_<chr>_` because of the "NA" string in the original Excel file. This variable could be more useful in our analysis if it was instead stored as logical. One option is to mutate the variable types on the existing object in our environment, or (taking the "go as far as possible at the import stage" approach) we can reread the contents of the file and apply the variable type definition at that point. The second option also gives us a way to explicitly define as `NA` those variables where it has been coded as a different character. Here `NA` is represented as "NA", but in other data you may encounter "not available", "9999", "-", and a host of other possibilities.

```
df_HR_train <-
  readxl::read_excel(
    dpjr::dpjr_data("cr25/cr25_human_resources.xlsx"),
    sheet = "df_HR_train",
    col_types = c("text", "logical", "logical", "logical", "logical"),
    na = "NA"
  )

glimpse(df_HR_train, width = 65)
```

```
## Rows: 999
## Columns: 5
## $ emp_id          <chr> "ID001", "ID002", "ID003", "ID004", "ID0~
## $ train_security <lgl> TRUE, FALSE, TRUE, FALSE, FALSE, FALSE, ~
## $ train_crm       <lgl> TRUE, FALSE, TRUE, TRUE, FALSE, TRUE, TR~
## $ train_qc        <lgl> NA, FALSE, FALSE, FALSE, FALSE, NA, FALS~
## $ train_history  <lgl> TRUE, TRUE, TRUE, TRUE, TRUE, TRUE, TRUE~
```

Now that the variables are all in a type that suits what is being represented, applying R's `summary()` function gives us another way to see the data.

```
summary(df_HR_train)
```

```
##     emp_id           train_security  train_crm        train_qc
##  Length:999         Mode :logical   Mode :logical   Mode :logical
##  Class :character   FALSE:500       FALSE:219       FALSE:494
##  Mode  :character   TRUE :462       TRUE :771       TRUE :110
##                     NA's :37        NA's :9         NA's :395
##  train_history
##  Mode :logical
##  FALSE:16
```

```
##  TRUE :971
##  NA's :12
```

This output is our preliminary EDA. What do we notice? Because there are no numeric values, we are not looking for outlier values or sentinel values or the range of the values.

A few NA values might be expected in a survey dataframe like this, and a handful may not have an impact on our analysis. But the `train_qc` variable has 395 missing values—nearly 40% of the 999 records. There might be a reason for a large number of missing values, and part of the analyst's responsibility is to understand why this might be the case. First, record that you've observed this in your notes about the data.

A thorough understanding of the data and its origins will ensure that your analysis is accurate. Further investigation (and recordkeeping) of this situation is warranted. You might start with the data documentation you received with the file. If that doesn't hold an explanation, talk to the people who understand how the data are collected, and then the people in the human resources department—perhaps there's a policy that exempts a group of employees from what is otherwise a mandatory course.

11.2.1 Data visualization

A plot can show us a wealth of information that cannot be expressed in the summary statistics. Let's work through exploring the date of birth variable from the `df_HR_main` sheet.

```
df_HR_main <-
  read_excel(
    path = dpjr::dpjr_data("cr25/cr25_human_resources.xlsx"),
    sheet = "df_HR_main",
    range = "B5:E1004"
  )

head(df_HR_main)
```

```
## # A tibble: 6 x 4
##   emp_id date_of_birth       name               gender
##   <chr>  <dttm>              <chr>              <chr>
## 1 ID001  1981-05-27 00:00:00 Avril Lavigne      Female
## 2 ID002  1973-09-24 00:00:00 Katy Perry         Female
## 3 ID003  1966-11-29 00:00:00 Chester Bennington Male
## 4 ID004  1982-09-17 00:00:00 Chris Cornell      Male
```

```
## 5 ID005  1976-07-17 00:00:00 Bryan Adams        Male
## 6 ID006  1965-07-03 00:00:00 Courtney Love      Female
```

Again we will use the `glimpse()` and `summary()` functions to start our exploration:

```
glimpse(df_HR_main)
```

```
## Rows: 999
## Columns: 4
## $ emp_id        <chr> "ID001", "ID002", "ID003", "ID004", "ID005", "ID006~
## $ date_of_birth <dttm> 1981-05-27, 1973-09-24, 1966-11-29, 1982-09-
17, 19~
## $ name         <chr> "Avril Lavigne", "Katy Perry", "Chester Bennington"~
## $ gender        <chr> "Female", "Female", "Male", "Male", "Male", "Female~
```

```
summary(df_HR_main)
```

```
##     emp_id            date_of_birth                          name
##  Length:999         Min.   :1901-01-01 00:00:00.000   Length:999
##  Class :character   1st Qu.:1965-10-25 12:00:00.000   Class :character
##  Mode  :character   Median :1976-06-21 00:00:00.000   Mode  :character
##                     Mean   :1976-01-14 17:53:52.432
##                     3rd Qu.:1987-03-08 12:00:00.000
##                     Max.   :1999-01-01 00:00:00.000
##     gender
##  Length:999
##  Class :character
##  Mode  :character
##
##
##
```

In the midst of this information is an interesting anomaly—you may have spotted it. It is one that is far more visible if we plot the data. Below is a density plot that shows the distribution of birth dates of the staff:

```
ggplot(df_HR_main, aes(x = date_of_birth )) +
  geom_density()
```

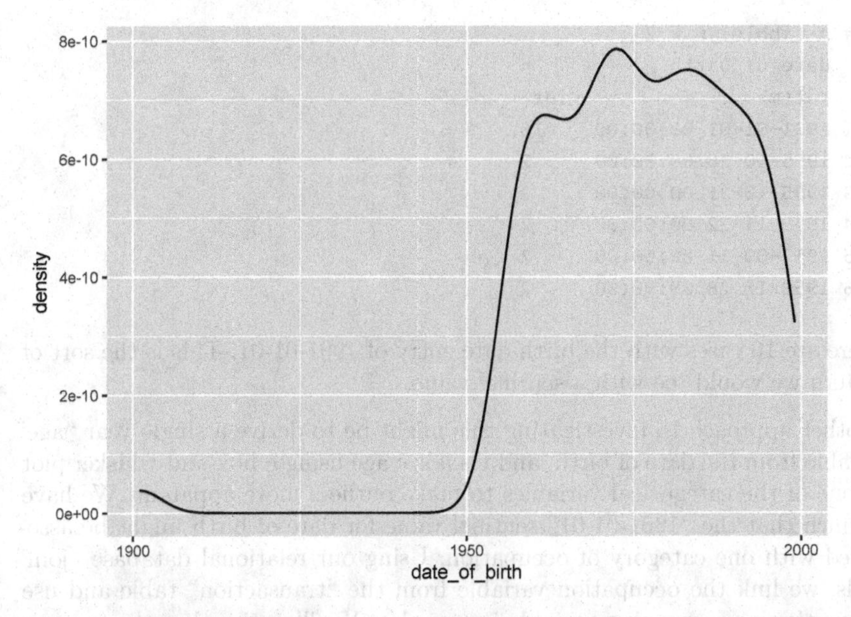

With a plot it is easy to see that there are a handful of birth dates around the year 1900, and then it takes off in the 1950s.

Here are some questions that we might want to explore further in the `date_of_birth` variable, after seeing the results of the `summary()` function and the plot:

- The earliest birth date is 1901-01-01. This would mean that the oldest staff member was 119 years old at 2020-01-01, the point at which the dataset was created. Perhaps you are already aware of the fact that only one person whose age has been independently verified lived long enough to celebrate their 120th birthday (they were 122 years, 164 days old when they passed away). Wikipedia, "List of the verified oldest people"[1]. A possibility is that 1901-01-01 is a sentinel value, inserted as a placeholder for unknown or missing values. Further exploration will help our understanding; we can start with counting the number of cases for any particular date:

```
df_HR_main |>
  group_by(date_of_birth) |>
  tally() |>
  arrange(desc(n)) |>
  head()
```

[1]https://en.wikipedia.org/wiki/List_of_the_verified_oldest_people

```
## # A tibble: 6 x 2
##   date_of_birth           n
##   <dttm>              <int>
## 1 1901-01-01 00:00:00    10
## 2 1955-10-26 00:00:00     2
## 3 1955-10-31 00:00:00     2
## 4 1955-11-22 00:00:00     2
## 5 1957-09-11 00:00:00     2
## 6 1958-10-28 00:00:00     2
```

There are 10 cases with the birth date entry of 1901-01-01. This is the sort of pattern we would see with a sentinel value.

Another approach to investigating this might be to derive a single year "age" variable from the date of birth, and visualize age using a box-and-whisker plot by one of the categorical variables to make outliers more apparent. We have a hunch that the "1901-01-01" sentinel value for date of birth might be associated with one category of occupation. Using our relational database "join" skills, we link the occupation variable from the "transaction" table and use occupation as a grouping variable in our plot. You'll notice that the transaction table is not in a structure that suits our analysis needs, so we have to do some data wrangling first.

```
# step 1: read transaction table, which contains occupation
df_HR_transaction <-
  read_excel(
    path = dpjr::dpjr_data("cr25/cr25_human_resources.xlsx"),
    sheet = "df_HR_transaction",
    na = "NA"
  )

# step 2: determine occupation on the date the table was created
df_HR_occupation_2020 <- df_HR_transaction |>
  group_by(emp_id) |>
  summarize(occupation_current = max(occupation, na.rm = TRUE))

# step 3: in the main table calculate age and
#   join the occupation table created in step 2,
#   then plot the result
df_HR_main |>
  # calculate age on January 1, 2020
  mutate(age_2020 =
           date_of_birth %--% ymd("2020-01-01") %/% years(1)) |>
  # join the current occupation table
```

```
left_join(df_HR_occupation_2020, by = "emp_id") |>
# plot
ggplot(aes(x = occupation_current, y = age_2020)) +
geom_boxplot()
```

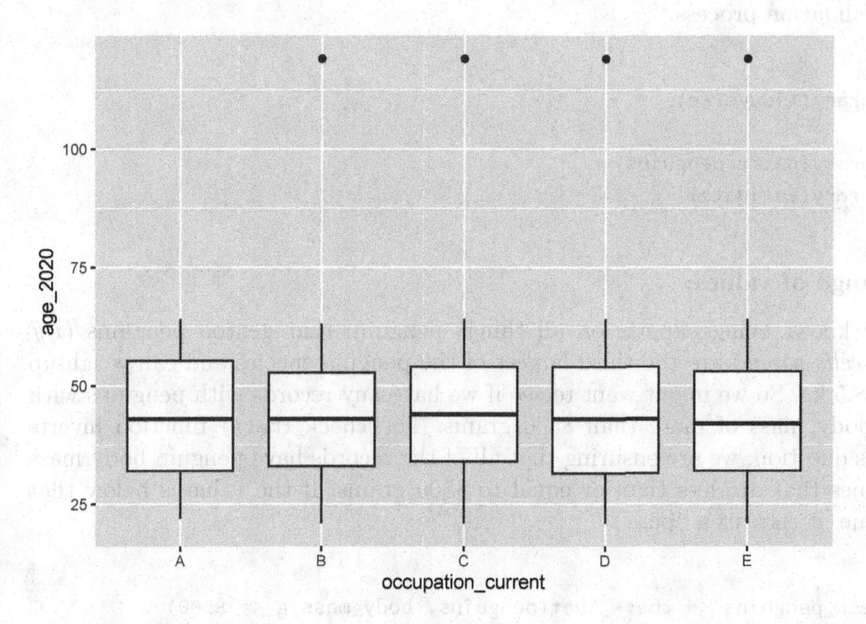

In the box-and-whisker visualization, we see that the sentinel values are in four of the five occupations, so the hunch was not correct.

11.3 Validation methods

While exploratory data analysis uses visualization and scans of data summaries to spot potentially dirty data, validation is a structured process of testing.

11.3.1 The {validate} package

As with other data analysis topics, people in the R user community have confronted the problem of dirty data and have written and shared functions to help the rest of us. The {validate} package (van der Loo and de Jonge, 2021, van der Loo et al. (2022)) was written and is supported by methodologists at

Statistics Netherlands, and provides tools for us to write structured validation tests.

The first function we will investigate in the package is `check_that()`, using the data in {palmerpenguins} (Horst et al., 2022; Horst, 2020). We know that the data in the package are nice and clean—they have already gone through a validation process.

```
library(tidyverse)

library(palmerpenguins)
library(validate)
```

Range of values:

We know, being experts on all things penguin, that gentoo penguins (*Pygoscelis papua*) are the third largest of the penguin species and can weigh up to 8.5 kg. So we might want to see if we have any records with penguins with a body mass of more than 8500 grams. The `check_that()` function inverts this question; we are ensuring that all of the records have penguin body mass values that are less than or equal to 8500 grams. If the value is below that value, it returns a "pass".

```
check_penguins <- check_that(penguins, body_mass_g <= 8500)

summary(check_penguins)
```

The `summary()` function on the object that the `check_that()` function creates shows that all of the penguins have a body mass of less than or equal to 8500 grams, and there are 2 "NA" values in the data.

Relationship:

While the heaviest gentoo penguins can be up to 8.5 kg, it is very unusual to find an Adélie or chinstrap penguin that weighs more than 5 kg. What happens if we change the body mass threshold to 5000 grams?

```
check_penguins <- check_that(penguins, body_mass_g <= 5000)

summary(check_penguins)
```

In addition to telling us that there are 61 records where the value in the `body_mass_g` variable is greater than 5000 grams, it is also letting us know that there are 2 `NA` values.

We can add an `if()` statement that filters out the gentoo penguins, and then check the body mass.

```
check_penguins <- check_that(
  penguins,
  if (species != "Gentoo") body_mass_g <= 5000
  )

summary(check_penguins)
```

Now all the records (except the single "NA") pass. What has happened? The first thing is that we have excluded all of the gentoo penguins—if it's gentoo, it gets marked "pass". All the remaining penguins (that is, the chinstrap and Adélie birds) are evaluated against the body mass value of 5000 grams. And they all pass.

Let's explore the data a bit more. If we filter our penguins by those that are over 5000 grams, what do we find?

```
penguins |>
  filter(body_mass_g > 5000) |>
  distinct(species)
```

```
## # A tibble: 1 x 1
##   species
##   <fct>
## 1 Gentoo
```

So this is confirms that all of the heavier penguins are gentoos, meeting our expectations.

Range:

We know that the {palmerpenguin} data is from three seasons of measurement, 2007–2009. We can write a validation test to ensure that our `year` variable falls within that range.

```
check_penguins <- check_that(penguins, year >= 2007 & year <= 2009)
summary(check_penguins)
```

```
##   name items passes fails nNA error warning
## 1   V1   344    344     0   0 FALSE   FALSE
##                                         expression
## 1 year - 2007 >= -1e-08 & year - 2009 <= 1e-08
```

What if we check for a single year?

```
check_penguins <- check_that(penguins, year == 2009)
summary(check_penguins)
```

```
##   name items passes fails nNA error warning                   expression
## 1   V1   344    120   224   0 FALSE   FALSE abs(year - 2009) <= 1e-08
```

The output gives us the number of cases in that year in the column `passes`, and the number of cases that are not in that year in the column `fails`.

Relationship:

We know that some species have only been measured on one of the three islands. Gentoo penguins have only been sampled on Biscoe Island, and chinstrap penguins in our sample come only from Dream Island. Does the data reflect that?

In the first test (labelled `g` for gentoo), if it's not a gentoo it gets included in the `passes`—and if it is a gentoo, it has to be on Biscoe Island to get a pass. In the `c` (for chinstrap) test, we will do the same for chinstrap penguins, which should all be on Dream Island.

```
check_penguins <-
  check_that(penguins,
             g = if (species == "Gentoo") island == "Biscoe",
             c = if (species == "Chinstrap") island == "Dream")

summary(check_penguins)
```

```
##   name items passes fails nNA error warning
## 1    g   344    344     0   0 FALSE   FALSE
## 2    c   344    344     0   0 FALSE   FALSE
##                                         expression
## 1   species != "Gentoo" | (island == "Biscoe")
## 2 species != "Chinstrap" | (island == "Dream")
```

The output now includes the results of the two tests. Both tests have all of the cases pass, and none fail.

We could have written code to generate a summary table to check this. In many datasets, however, there will be too many combinations to reliably investigate in this manner.

```
penguins |>
  group_by(species, island) |>
  tally()
```

```
## # A tibble: 5 x 3
## # Groups:   species [3]
##   species   island        n
##   <fct>     <fct>     <int>
## 1 Adelie    Biscoe       44
## 2 Adelie    Dream        56
## 3 Adelie    Torgersen    52
## 4 Chinstrap Dream        68
## 5 Gentoo    Biscoe      124
```

A single `check_that()` test could be written using the "AND" operator `&`.

```
check_penguins <-
  check_that(penguins,
             if (species == "Gentoo") island == "Biscoe" &
             if (species == "Chinstrap") island == "Dream")
```

```
summary(check_penguins)
```

```
  name items passes fails nNA error warning
1   V1   344    344     0   0 FALSE   FALSE
                                                                                expression
1 species != "Gentoo" | (island == "Biscoe" & (species != "Chinstrap" | (island == "Dream")))
```

FIGURE 11.1 *check_penguins.*

We can also test against a defined list. This strategy can be useful if there is an extensive list against which we need to evaluate our data. Imagine checking the spellings of every city and town in the region where you live. If you have a pre-defined list of those names in an existing data file, that list can be read into your R environment, and the test run against that list. In this short example, an object called `island_list` is created, and the validation check is run against the contents of that list.

```
island_list <- c("Biscoe", "Dream", "Torgersen")

check_penguins <-
  check_that(penguins,
             if (species == "Adelie") island %in% island_list)

summary(check_penguins)
```

```
##   name items passes fails nNA error warning
## 1   V1   344    344     0   0 FALSE   FALSE
##                                        expression
## 1 species != "Adelie" | (island %vin% island_list)
```

What happens if we fail to include Dream Island in our list of islands where Adélie penguins are found?

The {validator} also allows us to check variable types within our `check_that()` function. Because R stores a dataframe column as a single type, this returns only one pass/fail evaluation.

```
summary(check_that(penguins, is.integer(year)))
```

Description: df [1 × 8]

name <chr>	items <int>	passes <int>	fails <int>	nNA <int>	error <lgl>	warning <lgl>	expression <chr>
V1	1	1	0	0	FALSE	FALSE	is.integer(year)

1 row

FIGURE 11.2 *summary(check_that(penguins)).*

11.3.2 badpenguins

The "badpenguins.rds" file adds five fictional penguins to the original "penguins" table, including measurement data. Let's load the data file "badpenguins.rds" (an R data file) and run a couple of the tests we used earlier:

```
badpenguins <- read_rds(dpjr_data("badpenguins.rds"))
```

```
tail(badpenguins)
```

```
## # A tibble: 6 x 8
##   species island bill_length_mm bill_depth_mm flipper_length_mm
##   body_mass_g
##   <fct>   <fct>  <chr>          <chr>         <chr>          <chr>
## 1 Chinst~ Dream  50.2           18.7          198             3775
## 2 Penguin Gotham <NA>           <NA>          <NA>              90
## 3 Pingu   Antar~ <NA>           <NA>          <NA>            <NA>
## 4 Pinga   Antar~ <NA>           <NA>          <NA>            <NA>
## 5 Opus    Bloom~ <NA>           <NA>          <NA>            <NA>
## 6 Gunter  Ice K~ <NA>           <NA>          <NA>            <NA>
## # i 2 more variables: sex <fct>, year <chr>
```

```
check_penguins <- check_that(badpenguins, year >= 2007 & year <= 2009)

summary(check_penguins)
```

```
##   name items passes fails nNA error warning                   expression
## 1   V1   349    344     5   0 FALSE   FALSE year >= 2007 & year <= 2009
```

```
check_penguins <- check_that(badpenguins, island %in% island_list)

summary(check_penguins)
```

```
##   name items passes fails nNA error warning                  expression
## 1   V1   349    344     5   0 FALSE   FALSE island %vin% island_list
```

In both cases, we get 5 "fails".

11.3.3 Investigating the fails

How can we check which records have failed?

{validate} provides two other functions, `validator()` and `confront()`, which give us a way to

- run multiple checks at a time, and

- generate a detailed record-by-record account of which records have failed our test.

First, we assign the syntax of our tests using `validator()` to an object "penguin_tests".

```
penguin_tests <- validator(
  island %in% island_list,
  year >= 2007 & year <= 2009,
  body_mass_g <= 8500
)

penguin_tests
```

```
## Object of class 'validator' with 3 elements:
##  V1: island %in% island_list
##  V2: year >= 2007 & year <= 2009
##  V3: body_mass_g <= 8500
```

In the next step, we apply those tests in a sequence of "confrontations" using the `confront()` function, generating an output object that we can investigate.

```
penguin_confront <- confront(badpenguins, penguin_tests)

penguin_confront
```

```
## Object of class 'validation'
## Call:
##     confront(dat = badpenguins, x = penguin_tests)
##
## Rules confronted: 3
##    With fails    : 3
##    With missings: 1
##    Threw warning: 0
##    Threw error  : 0
```

```
summary(penguin_confront)
```

```
##   name items passes fails nNA error warning                        expression
## 1   V1   349    344     5   0 FALSE   FALSE    island %vin% island_list
## 2   V2   349    344     5   0 FALSE   FALSE year >= 2007 & year <= 2009
## 3   V3   349    342     1   6 FALSE   FALSE              body_mass_g <= 8500
```

Let's look at the last ten rows of the results of the `confront()` function of our `badpenguins` data. Rows 340 through 344 are original to the clean `penguins` dataset; 345 through 349 are the bad penguins. The first of the bad penguins fails on all three tests, while the others fail on the first two and have "NA" values for the third.

```
tail(values(penguin_confront), 10)
```

```
##            V1    V2    V3
## [340,]  TRUE  TRUE  TRUE
## [341,]  TRUE  TRUE  TRUE
## [342,]  TRUE  TRUE  TRUE
## [343,]  TRUE  TRUE  TRUE
## [344,]  TRUE  TRUE  TRUE
## [345,] FALSE FALSE FALSE
## [346,] FALSE FALSE    NA
## [347,] FALSE FALSE    NA
## [348,] FALSE FALSE    NA
## [349,] FALSE FALSE    NA
```

In this example, our prior knowledge of the data gives us the insight to quickly target the rows that failed the validation tests. In the real world, filtering the object that results from our validation tests will allow us to identify the rows that contain the records that fail the tests.

The function we use for this is `violating()`. The two arguments in the `violating()` function are

- the original data frame `badpenguins` and
- the object that resulted from the `confront()` function, `penguin_confront`.

```
violating(badpenguins, penguin_confront)
```

A tibble: 5 x 8

species <fctr>	island <fctr>	bill_length_mm <chr>	bill_depth_mm <chr>	flipper_length_mm <chr>	body_mass_g <chr>	sex <fctr>	year <chr>
Penguin	Gotham	NA	NA	NA	90	male	1966
Pingu	Antarctica	NA	NA	NA	NA	male	1986
Pinga	Antarctica	NA	NA	NA	NA	female	1986
Opus	Bloom County	NA	NA	NA	NA	male	1981
Gunter	Ice Kingdom	NA	NA	NA	NA	unknown	2012

5 rows

FIGURE 11.3 *violating(bad_penguins).*

Now that we know the details of the cases that have failed the test, we can make some decisions—based on our understanding of the research we are undertaking—about how to deal with these cases.

12

Cleaning techniques

In this chapter:

- Cleaning dates and strings
- Creating conditional and calculated variables
  - indicator variables (also known as dummy variables (Suits, 1957) and "one hot" encoding)
- Dealing with missing values

12.1 Introduction

We have already applied some cleaning methods. In the chapters on importing data, we tackled a variety of techniques that would fall into "data cleaning," including assigning variable types and restructuring data into a tidy format. In this chapter, we look in more depth at some examples of recurring data cleaning challenges.

12.2 Cleaning dates

A recurring challenge in many data analysis projects is to deal with date and time fields. One of the first steps of any data cleaning project should be to ensure that these are in a consistent format that can be easily manipulated.

As we saw in (Broman and Woo, 2017), there is an international standard (ISO 8601) for an unambiguous and principled way to write dates and times (and in combination, date-times). For dates, we write **YYYY-MM-DD**. Similarly, time is written **T[hh][mm][ss]** or **T[hh]:[mm]:[ss]**, with hours (on a 24-hour clock) followed by minutes and seconds.

When we are working with dates, the objective of our cleaning is to transform the date into the ISO 8601 standard and then, where appropriate, transform that structure into a date type variable. The benefits of the date type format are many, not the least of which are the R tools built to streamline analysis and visualization that use date type variables, (Wickham et al., 2023b, Chapter 18) and those same tools allow us to tackle cleaning variables that capture date elements.

One such package is {lubridate}, part of the tidyverse.

12.2.1 Creating a date-time field

The first way that dates might be represented in your data is as three separate fields, one for year, one for month, and one for day. These can be assembled into a single date variable using the `make_date()`, or a single datetime variable using `make_datetime()`.

In the {Lahman}(Friendly et al., 2020) table "People", each player's birth date is stored as three separate variables. They can be combined into a single variable using `make_date()`:

```
Lahman::People |>
  slice_head(n = 10) |>
  select(playerID, birthYear, birthMonth, birthMonth, birthDay) |>
  # create date-time variable
   mutate(birthDate = make_datetime(birthYear, birthMonth, birthDay))
```

```
##     playerID birthYear birthMonth birthDay  birthDate
## 1  aardsda01      1981         12       27 1981-12-27
## 2  aaronha01      1934          2        5 1934-02-05
## 3  aaronto01      1939          8        5 1939-08-05
## 4   aasedo01      1954          9        8 1954-09-08
## 5   abadan01      1972          8       25 1972-08-25
## 6   abadfe01      1985         12       17 1985-12-17
## 7  abadijo01      1850         11        4 1850-11-04
## 8  abbated01      1877          4       15 1877-04-15
## 9  abbeybe01      1869         11       11 1869-11-11
## 10 abbeych01      1866         10       14 1866-10-14
```

The second problematic way dates are stored is in a variety of often ambiguous sequences. This might included text strings for the month (for example "Jan",

"January", or "Janvier" rather than "01"), two-digit years (does "12" refer to 2012 or 1912?), and ambiguous ordering of months and days (does "12" refer to the twelfth day of the month or December?)

If we know the way the data is stored, we can use {lubridate} functions to process the data into YYYY-MM-DD. These three very different date formats can be transformed into a consistent format:

```
ymd("1973-10-22")
```

```
## [1] "1973-10-22"
```

```
mdy("October 22nd, 1973")
```

```
## [1] "1973-10-22"
```

```
dmy("22-Oct-1973")
```

```
## [1] "1973-10-22"
```

A third manner in which dates can be stored is with a truncated representation. For example, total sales for the entire month might be recorded as a single value associated with "Jan 2017" or "2017-01", or a quarterly measurement might be "2017-Q1".

```
# month
ym("1973-10")
```

```
## [1] "1973-10-01"
```

```
my("Oct 1973")
```

```
## [1] "1973-10-01"
```

```
# quarter
yq("1973-Q1")
```

```
## [1] "1973-01-01"
```

If we have a variable with annual data, we have to use the parameter `truncated =` to convert our year into a date format:

```
ymd("1973", truncated = 2)
```

```
## [1] "1973-01-01"
```

By mutating these various representations into a consistent YYYY-MM-DD format, we can now run calculations. In the example below, we calculate the average size of the Canadian labour force in a given year:

```
lfs_canada_employment <-
  read_rds(dpjr::dpjr_data("lfs_canada_employment.rds"))
```

First, we transform the character variable `REF_DATE` into a date type variable.

```
lfs_canada_employment <- lfs_canada_employment |>
  mutate(REF_DATE = ym(REF_DATE))
```

With the variable in the date type, we can use functions within {lubridate} for our analysis.

To calculate the average employment in each year:

```
lfs_canada_employment |>
  group_by(year(REF_DATE)) |>
  summarize(mean(VALUE)) |>
  slice_head(n = 4)
```

```
## # A tibble: 4 x 2
##   `year(REF_DATE)` `mean(VALUE)`
##              <dbl>         <dbl>
## 1             2011        17244.
## 2             2012        17476.
## 3             2013        17712.
## 4             2014        17783.
```

For the second analysis, we might be interested in determining if there is any seasonality in the Canadian labour force:

```
lfs_canada_employment |>
  group_by(month(REF_DATE)) |>
  summarize(mean(VALUE))
```

```
## # A tibble: 12 x 2
##    `month(REF_DATE)` `mean(VALUE)`
##                <dbl>         <dbl>
##  1                 1        17841.
##  2                 2        17960.
##  3                 3        17907.
##  4                 4        17848.
##  5                 5        18255.
##  6                 6        18539.
##  7                 7        18514.
##  8                 8        18514.
##  9                 9        18423.
## 10                10        18458.
## 11                11        18421.
## 12                12        18369.
```

12.3 Cleaning strings

"Strings are not glamorous, high-profile components of R, but they do play a big role in many data cleaning and preparation tasks." —{stringr} reference page

R provides some robust tools for working with character strings. In particular, the {stringr}(Wickham, 2019b) package is very flexible. But before we get there, we need to take a look at regular expressions.

12.3.1 Regular expressions

"Some people, when confronted with a problem, think: 'I know, I'll use regular expressions.' Now they have two problems." — Jamie Zawinski

Regular expressions are a way to specify a pattern of characters. This pattern can be specific letters or numbers ("abc" or "123") or character types (letters, digits, white space, etc.), or a combination of the two.

An excellent resource for learning the basics is the vignette "Regular Expressions," hosted at the {stringr} package website "Regular Expressions", stringr.tidyverse.org[1].

Here are some useful regex matching functions. Note that the square brackets enclose the regex to identify a single character. In the examples below, with three sets of square brackets, "[a][b][c]" contains three separate character positions, whereas "[abc]" is identifying a single character.

character	what it does
"abc"	matches "abc"
"[a][b][c]"	matches "abc"
"[abc]"	matches "a", "b", or "c"
"[a-c]"	matches any of "a" to "c" (that is, matches "a", "b", or "c")
"[^abc]"	matches anything *except* "a", "b", or "c"
"^"	match start of string
"$"	match end of string
"."	matches any string

character	frequency of match
"?"	0 or 1
"+"	1 or more
"*"	0 or more
"{n}"	exactly n times
"{n,}"	n or more
"{n,m}"	between n and m times

These also need to be escaped:

character	what it is
"""	double quotation mark
"'"	single quotation mark
"\"	backslash
"\d"	any digit
"\n"	newline (line break)
"\s"	any whitespace (space, tab, newline)
"\t"	tab
"\u..."	unicode characters*

[1]https://stringr.tidyverse.org/articles/regular-expressions.html

An example of a unicode character[2] is the "interobang", a combination of question mark and exclamation mark.

```
interrobang <- "\u2048"
interrobang
```

```
## [1] "?!"
```

Canadian postal codes follow a consistent pattern:

- letter digit letter, followed by a space or a hyphen, then digit letter digit

The regex for this is shown below, with the component parts as follows:

- ^ : start of string
- [A-Za-z] : all the letters, both upper and lower case
- \d : any numerical digit
- [-]? : space or hyphen; "?" to make either optional (i.e. it may be there or not)
- $: end of string

```
# regex for Canadian postal codes
# ^ : start of string
# [A-Za-z] : all the letters upper and lower case
# \\d : any numerical digit
# [ -]? : space or hyphen,
#          ? to make either optional (i.e. it may be there or not)
# $ : end of string

postalcode <- "^[A-Za-z]\\d[A-Za-z][ -]?\\d[A-Za-z]\\d$"
```

> "If your regular expression gets overly complicated, try breaking it up into smaller pieces, giving each piece a name, and then combining the pieces with logical operations." —Wickham

[2]The full list of unicode characters can be found at the Wikipedia webpage (Wikipedia contributors, 2022a).

and Grolemund, *R for Data Science* (Wickham and Grolemund, 2016, p.209)

The regular expression for Canadian postal codes can be shortened, using the case sensitivity operator, "(?i)". Putting this at the beginning of the expression applies it to all of the letter identifiers in the regex; it can be turned off with "(?-i)".

```
# option: (?i) : make it case insensitive
#         (?-i) : turn off insensitivity
postalcode <- "(?i)^[A-Z]\\d[A-Z][ -]?\\d[A-Z]\\d$(?-i)"
```

The postal code for the town of Kimberley, British Columbia, is "V1A 2B3". Here's a list with some variations of that code, which we will use to test our regular expression.

```
# list of postal codes
pc_kimberley <- c("V1A2B3", "V1A 2B3", "v1a 2B3xyz", "v1a-2b3")
```

12.3.2 Data cleaning with {stringr}

Armed with the power of regular expressions, we can use the functions within the {stringr} package to filter, select, and clean our data.

```
library(stringr)
```

As with many R packages, the reference page provides good explanations of the functions in the package, as well as vignettes that give working examples of those functions. The reference page for {stringr} is stringr.tidyverse.org[3].

Let's look at some of the functions within {stringr} that are particularly suited to cleaning tasks.

[3]https://stringr.tidyverse.org/

12.3.2.1 String characteristics

function	purpose
`str_subset()`	character; returns the cases which contain a match
`str_which()`	numeric; returns the vector location of matches
`str_detect()`	logical; returns TRUE if pattern is found

The {stringr} functions `str_subset()` and `str_which()` return the strings that have the pattern, either in whole or by location.

```
# which strings have the postal code pattern?
str_subset(pc_kimberley, postalcode)
```

```
## [1] "V1A2B3"  "V1A 2B3" "v1a-2b3"
```

```
# which strings have the postal code pattern, by location?
str_which(pc_kimberley, postalcode)
```

```
## [1] 1 2 4
```

Another of the {stringr} function, `str_detect()`, tests for the presence of a pattern and returns a logical value (true or false). Here, we test the list of four postal codes using the `postalcode` object that contains the regular expression for Canadian postal codes.

```
str_detect(string = pc_kimberley, pattern = postalcode)
```

```
## [1]  TRUE  TRUE FALSE  TRUE
```

As we see, the third example postal code—which contains an extra digit at the end—returns a "FALSE" value. It might contain something that looks like a postal code in the first 7 spaces, but because it has the extra letter, it is not a postal code. To find a string that matches the postal code in any part of a string, we have to change our regex to remove the start "^" and end "$" specifications:

```
# remove start and end specifications
postalcode <- "(?i)[A-Z]\\d[A-Z][ -]?\\d[A-Z]\\d(?-i)"
```

```
str_detect(string = pc_kimberley, pattern = postalcode)
```

```
## [1] TRUE TRUE TRUE TRUE
```

All of the strings have something that matches the structure of a Canadian postal code. How can we extract the parts that match?

If we rerun the `str_subset()` and `str_which()` functions, we see that all four of the strings are now returned.

```
# which strings have the pattern?
str_subset(pc_kimberley, postalcode)
```

```
## [1] "V1A2B3"     "V1A 2B3"    "v1a 2B3xyz" "v1a-2b3"
```

```
# which strings have the pattern ("e"), by location?
str_which(pc_kimberley, postalcode)
```

```
## [1] 1 2 3 4
```

With those two functions, we can identify which strings have the characters that we're looking for. To retrieve a particular string, we can use `str_extract()`:

```
pc_kimberley_all <- str_extract(pc_kimberley, postalcode)
pc_kimberley_all
```

```
## [1] "V1A2B3"  "V1A 2B3" "v1a 2B3" "v1a-2b3"
```

12.3.2.2 Replacing and splitting functions

How can we clean up this list so that our postal codes are in a consistent format?

Our goal will be a string of six characters, with upper-case letters and neither a space nor a hyphen.

This table shows the functions we will use.

function	purpose
`str_match()`	extracts the text of the match and parts of the match within parenthesis
`str_replace()` and `str_replace_all()`	replaces the matches with new text
`str_remove()` and `str_remove_all()`	removes the matches
`str_split()`	splits up a string at the pattern

The first step will be to remove the space where the pattern is `" "` using the `stringr::str_remove()` function. This step would then be repeated with `pattern = "-"`

```
str_remove(string = pc_kimberley_all,
           pattern = " ")
```

```
## [1] "V1A2B3"  "V1A2B3"  "v1a2B3"  "v1a-2b3"
```

More efficiently, we can use a regular expression with both a space and a hyphen inside the square brackets, which has the result of replacing *either* a space or a hyphen.

```
pc_kimberley_clean <- str_remove(string = pc_kimberley_all,
           pattern = "[ -]")
```

The second step will be to capitalize the letters. The {stringr} function for this is `str_to_upper`. (There are also parallel `str_to_lower`, `str_to_title`, and `str_to_sentence` functions.)

```
pc_kimberley_clean <- str_to_upper(pc_kimberley_clean)
pc_kimberley_clean
```

```
## [1] "V1A2B3" "V1A2B3" "V1A2B3" "V1A2B3"
```

If this was in a dataframe, these functions can be added to a `mutate()` function, and applied in a pipe sequence, as shown below.

```
# convert string to tibble
pc_kimberley_tbl <- as_tibble(pc_kimberley)
```

```
# mutate new values
# --note that creating three separate variables allows for comparisons
#   as the values change from one step to the next
pc_kimberley_tbl |>
  # extract the postal codes
  mutate(pc_extract = str_extract(value, postalcode)) |>
  # remove space or hyphen
  mutate(pc_remove = str_remove(pc_extract, "[ -]")) |>
  # change to upper case
  mutate(pc_upper = str_to_upper(pc_remove))
```

```
## # A tibble: 4 x 4
##   value      pc_extract pc_remove pc_upper
##   <chr>      <chr>      <chr>     <chr>
## 1 V1A2B3     V1A2B3     V1A2B3    V1A2B3
## 2 V1A 2B3    V1A 2B3    V1A2B3    V1A2B3
## 3 v1a 2B3xyz v1a 2B3    v1a2B3    V1A2B3
## 4 v1a-2b3    v1a-2b3    v1a2b3    V1A2B3
```

12.3.2.3 Split and combine strings

Below, we split an address based on comma and space location, using the `str_split()` function.

```
UVic_address <-
  "Continuing Studies,
   3800 Finnerty Rd,
   Victoria, BC,
   V8P 5C2"

str_split(UVic_address, ", ")
```

```
## [[1]]
## [1] "Continuing Studies,\n   3800 Finnerty Rd"
## [2] "\n   Victoria"
## [3] "BC"
## [4] "\n   V8P 5C2"
```

{stringr} contains three useful functions for combining strings. `str_c()` collapses multiple strings into a single string, separated by whatever we specify (the default is nothing, that is to say, `""`.)

Below we will combine the components of an address into a single string, with the components separated by a comma and a space.

```
UVic_address_components <- c(
  "Continuing Studies",
   "3800 Finnerty Rd",
   "Victoria", "BC",
   "V8P 5C2"
  )

str_c(UVic_address_components, collapse = ", ")
```

```
## [1] "Continuing Studies, 3800 Finnerty Rd, Victoria, BC, V8P 5C2"
```

The `str_flatten_comma()` is a shortcut to the same result:

```
str_flatten_comma(UVic_address_components)
```

```
## [1] "Continuing Studies, 3800 Finnerty Rd, Victoria, BC, V8P 5C2"
```

Note that `str_flatten` could also be used, but we need to explicitly specify the separator:

```
str_flatten(UVic_address_components, collapse = ", ")
```

```
## [1] "Continuing Studies, 3800 Finnerty Rd, Victoria, BC, V8P 5C2"
```

These functions also have the argument `last =`, which allows us to specify the final separator:

```
str_flatten(UVic_address_components, collapse = ", ", last = " ")
```

```
## [1] "Continuing Studies, 3800 Finnerty Rd, Victoria, BC V8P 5C2"
```

12.3.3 Example: Bureau of Labor Statistics by NAICS code

In this example, we will use regular expressions to filter a table of employment data published by the US Bureau of Labor Statistics. The table shows the annual employment by industry, for the years 2010 to 2020, in thousands of people.

With this data, it's important to understand the structure of the file, which is rooted in the North American Industry Classification System (NAICS). This typology is used in Canada, the United States, and Mexico, grouping companies by their primary output. This consistency allows for reliable and comparable analysis of the economies of the three countries. More information can be found in (Executive Office of the President, Office of Management and Budget, 2022) and (Statistics Canada, 2022).

The NAICS system, and the file we are working with, is hierarchical; higher-level categories subdivide into subsets. Let's examine employment in the Mining, Quarrying, and Oil and Gas Extraction sector.

This sector's coding is as follows:

Industry	NAICS code
Mining, Quarrying, and Oil and Gas Extraction	21
* Oil and Gas Extraction	211
* Mining (except Oil and Gas)	212
- Coal Mining	2121
- Metal Ore Mining	2122
- Nonmetallic Mineral Mining and Quarrying	2123
* Support Activities for Mining	213

Let's look at the data now. The original CSV file has the data for 2010 through 2020; for this example the years 2019 and 2020 are selected. Note that the table is rendered using the {gt} package (Iannone et al., 2023).

```
bls_employment <-
  read_csv(dpjr::dpjr_data("us_bls_employment_2010-2020.csv")) |>
  select(`Series ID`, `Annual 2019`, `Annual 2020`)

gt::gt(head(bls_employment))
```

Series ID	Annual 2019	Annual 2020
CEU0000000001	150904.0	142186.0
CEU0500000001	128291.0	120200.0
CEU0600000001	21037.0	20023.0
CEU1000000001	727.0	600.0
CEU1011330001	49.0	46.4
CEU1021000001	678.1	553.0

There are three "supersectors" in the NAICS system: primary industries, goods-producing industries, and service-producing industries. The "Series ID" variable string starts with "CEU" followed by the digits in positions 4 and 5 representing the supersector. The first row, supersector "00", is the aggregation of the entire workforce.

Mining, Quarrying, and Oil and Gas Extraction is a sector within the goods-producing industries, with the NAICS code number 21. This is represented at digits 6 and 7 of the string.

Mining (NAICS 212) is a subsector, comprising three industries (at the four-digit level).

The hierarchical structure of the dataframe means that if we simply summed the column we would end up double-counting: the sum of coal, metal ore, and non-metallic mineral industry rows is reported as the employment in Mining (except Oil and Gas).

Let's imagine our assignment is to produce a chart showing the employment in the three subsectors, represented at the three-digit level.

First, let's create a table with all the rows associated with the sector, NAICS 21.

```
# create NAICS 21 table
bls_employment_naics21 <- bls_employment |>
  filter(str_detect(`Series ID`, "CEU1021"))

gt(
  bls_employment_naics21
)
```

Series ID	Annual 2019	Annual 2020
CEU1021000001	678.1	553.0
CEU1021100001	143.6	129.8
CEU1021200001	190.4	176.3
CEU1021210001	50.5	39.8
CEU1021220001	42.2	41.4
CEU1021230001	97.8	95.1
CEU1021300001	344.0	246.8

In the table above, we see the first row has the "Series ID" of CEU1021000001. That long string of zeros shows that this is the sector, NAICS 21. The employment total for each year is the aggregation of the subsectors 211, 212, and

213. As well, this table also shows the industries within 212, starting with CEU1021210001 (NAICS 2121).

What if we selected only the values that have a zero in the ninth spot?

```
gt(
bls_employment_naics21 |>
  # remove the ones that have a zero in the 5th spot from the end
  filter(str_detect(`Series ID`, "00001$"))
)
```

Series ID	Annual 2019	Annual 2020
CEU1021000001	678.1	553.0
CEU1021100001	143.6	129.8
CEU1021200001	190.4	176.3
CEU1021300001	344.0	246.8

That's not quite right—it includes the "21" sector aggregate.

Let's filter out the rows that have a zero after the "21" using an exclamation mark at the beginning of our `str_detect()` function.

```
# NAICS 212
bls_employment_naics21 |>
  # use the exclamation mark to filter those that don't match
  filter(!str_detect(`Series ID`, "CEU10210"))
```

```
## # A tibble: 6 x 3
##   `Series ID`   `Annual 2019` `Annual 2020`
##   <chr>                 <dbl>         <dbl>
## 1 CEU1021100001         144.          130.
## 2 CEU1021200001         190.          176.
## 3 CEU1021210001          50.5          39.8
## 4 CEU1021220001          42.2          41.4
## 5 CEU1021230001          97.8          95.1
## 6 CEU1021300001         344           247.
```

That doesn't work, because it leaves in the four-digit industries, starting with 2121.

The solution can be found by focusing on the characters that differentiate the levels in the hierarchy.

Let's add the digits 1 to 9 to the front of the second filter (the 8th character of the string), and leave the next character (the 9th character) as a zero. This should find "CEU1021200001" but omit "CEU1021210001".

```
# NAICS 212
bls_employment_naics21 |>
  # remove the ones that have a zero in the 5th spot from the end
  # - add any digit from 1-9, leave the following digit as 0
  filter(str_detect(`Series ID`, "[1-9]00001$"))
```

```
## # A tibble: 3 x 3
##   `Series ID`   `Annual 2019` `Annual 2020`
##   <chr>                 <dbl>         <dbl>
## 1 CEU1021100001          144.          130.
## 2 CEU1021200001          190.          176.
## 3 CEU1021300001          344           247.
```

A second option would be to use a negate zero, "[^0]". You'll notice that this uses the carat symbol, "^". Because it is inside the square brackets, it has the effect of negating the named values. This is different behaviour than when it's outside the square brackets and means "start of the string".

```
# NAICS 212
bls_employment_naics21 |>
  # remove the ones that have a zero in the 6th spot from the end
  # - add any digit that is not a zero
  filter(str_detect(`Series ID`, "[^0]00001$"))
```

```
## # A tibble: 3 x 3
##   `Series ID`   `Annual 2019` `Annual 2020`
##   <chr>                 <dbl>         <dbl>
## 1 CEU1021100001          144.          130.
## 2 CEU1021200001          190.          176.
## 3 CEU1021300001          344           247.
```

And here's one final option, using `str_sub()` to specify the location, inside the `str_detect()` function.

Let's look at the results of the `str_sub()` function in isolation before we move on. In this code, the function extracts all of the character strings from the 8th and 9th positions in the "Series ID" variable, resulting in strings that are two characters long.

```
# extract the character strings in the 8th & 9th position
str_sub(bls_employment_naics21$`Series ID`, start = 8, end = 9)
```

```
## [1] "00" "10" "20" "21" "22" "23" "30"
```

Starting on the inside of the parentheses, we have the `str_sub()` function to create those two-character strings. Next is the `str_detect()` with a regex that identifies those that have a digit other than zero in the first position, and a zero in the second. The table is then filtered for those rows where that was detected.

```
# NAICS 212
bls_employment_naics21_3digit <- bls_employment_naics21 |>
  # filter those that match the regex in the 8th & 9th position
  filter(str_detect(
    str_sub(`Series ID`, start = 8, end = 9),
  "[^0][0]"))

bls_employment_naics21_3digit
```

```
## # A tibble: 3 x 3
##   `Series ID`   `Annual 2019` `Annual 2020`
##   <chr>                 <dbl>         <dbl>
## 1 CEU1021100001          144.          130.
## 2 CEU1021200001          190.          176.
## 3 CEU1021300001          344           247.
```

The final step in cleaning this table would be to put it into a tidy structure. Currently it violates the second principle of tidy data, that each observation forms a row. (Wickham, 2014, p.4) In this case, each year should have its own row, rather than a separate column for each.

We can use the {tidyr} (Wickham, 2021b) package's `pivot_longer()` function for this:

```
bls_employment_naics21_3digit |>
  pivot_longer(
    cols = contains("Annual"),
    names_to = "year",
    values_to = "employment")
```

```
## # A tibble: 6 x 3
##   `Series ID`   year        employment
##   <chr>         <chr>            <dbl>
## 1 CEU1021100001 Annual 2019       144.
## 2 CEU1021100001 Annual 2020       130.
## 3 CEU1021200001 Annual 2019       190.
## 4 CEU1021200001 Annual 2020       176.
## 5 CEU1021300001 Annual 2019       344
## 6 CEU1021300001 Annual 2020       247.
```

12.3.4 Example: NOC: split code from title

This example is drawn from a table that shows the number of people working in different occupations in British Columbia's accommodation industry. In this table the Canadian National Occupation Classification (NOC) code is in the same column as the code's title.

```
noc_table <- read_rds(dpjr::dpjr_data("noc_table.rds"))

gt(head(noc_table))
```

noc_occupation_title	certification_training_requirements	employment
063: Accommodation service managers	□ A university degree or college diploma in hotel management or other related discipline or equivalent job experience	6,479
652: Occupations in travel and accommodation	□ Related post-secondary diploma; On the job training	5,771
673: Cleaners	□ On the job training	5,083
631: Food service supervisors	□ Post-secondary training in restaurant management or food service admin, or; Equivalent job experience	3,479
632: Chefs and cooks	□ Cook's apprenticeship program and training; Chef's Red Seal Certification	1,542
112: Human resources professionals	□ Related post-secondary degree or diploma	1,271

FIGURE 12.1 *NOC table: step 1.*

This table has four variables in three columns, each of which has at least one data cleaning challenge.

noc_occupation_title

This variable combines two values: the three-digit NOC code and the title of the occupation.

We will use the {tidyr} function `separate_wider_delim()` to split it into two variables. The function gives us the arguments to name the new columns that result from the separation and the character string that will be used to identify the separation point, in this case ": ". **Don't forget the space!**

This function also has an argument `cols_remove =`. The default is set to `TRUE`, which removes the original column. Set to `FALSE` it retains the original column; we won't use that here but it can be useful.

There are also `separate_wider_()` functions based on fixed-width positions and regular expressions.

```
noc_table_clean <- noc_table |>
  tidyr::separate_wider_delim(
    cols = noc_occupation_title,
    delim = ": ",
    names = c("noc", "occupation_title")
    )


gt(head(noc_table_clean))
```

noc	occupation_title	certification_training_requirements	employment
063	Accommodation service managers	□ A university degree or college diploma in hotel management or other related discipline or equivalent job experience	6,479
652	Occupations in travel and accommodation	□ Related post-secondary diploma; On the job training	5,771
673	Cleaners	□ On the job training	5,083
631	Food service supervisors	□ Post-secondary training in restaurant management or food service admin, or; Equivalent job experience	3,479
632	Chefs and cooks	□ Cook's apprenticeship program and training; Chef's Red Seal Certification	1,542
112	Human resources professionals	□ Related post-secondary degree or diploma	1,271

FIGURE 12.2 *NOC table: step 2.*

certification_training_requirements

This variable has a character that gets represented as a square at the beginning, and also a space between the square and the first letter.

Here we can use the {stringr} function `str_remove_all()`. (Remember that the `str_remove()` function removes only the first instance of the character.)

```
noc_table_clean <- noc_table_clean |>
  mutate(
    certification_training_requirements =
      str_remove_all(certification_training_requirements, "? ")
    )

gt(head(noc_table_clean))
```

noc	occupation_title	certification_training_requirements	employment
063	Accommodation service managers	A university degree or college diploma in hotel management or other related discipline or equivalent job experience	6,479
652	Occupations in travel and accommodation	Related post-secondary diploma; On the job training	5,771
673	Cleaners	On the job training	5,083
631	Food service supervisors	Post-secondary training in restaurant management or food service admin, or; Equivalent job experience	3,479
632	Chefs and cooks	Cook's apprenticeship program and training; Chef's Red Seal Certification	1,542
112	Human resources professionals	Related post-secondary degree or diploma	1,271

FIGURE 12.3 *NOC table: step 3.*

employment

This variable should be numbers but is a character type. This is because

- there are commas used as the thousands separator
- the smallest categories are represented with "-*" to indicate a table note.

In order to clean this, we require three steps:

1. remove commas

2. replace "-*" with NA
3. convert to numeric value

```
noc_table_clean <- noc_table_clean |>
  # remove commas
  mutate(
    employment =
      str_remove_all(employment, ",")
    ) |>
  # replace with NA
  mutate(
    employment =
      str_replace_all(employment, "-\\*", replacement = NA_character_)
    ) |>
  # convert to numeric
  mutate(
    employment =
      as.numeric(employment)
    )

gt(head(noc_table_clean))
```

noc	occupation_title	certification_training_requirements	employment
063	Accommodation service managers	A university degree or college diploma in hotel management or other related discipline or equivalent job experience	6479
652	Occupations in travel and accommodation	Related post-secondary diploma; On the job training	5771
673	Cleaners	On the job training	5083
631	Food service supervisors	Post-secondary training in restaurant management or food service admin, or; Equivalent job experience	3479
632	Chefs and cooks	Cook's apprenticeship program and training; Chef's Red Seal Certification	1542
112	Human resources professionals	Related post-secondary degree or diploma	1271

FIGURE 12.4 *NOC table: step 4.*

12.4 Creating conditional and calculated variables

We sometimes find ourselves in a situation where we want to recode our existing variables into other categories. An instance of this is when there are a handful of categories that make up the bulk of the total, and summarizing the smaller categories into an "all other" makes the table or plot easier to read. One example of this are the populations of the 13 Canadian provinces and territories: the four most populous provinces (Ontario, Quebec, British Columbia, and Alberta) account for roughly 85% of the country's population...the other 15% of Canadians are spread across nine provinces and territories. We might want to show a table with only five rows, with the smallest nine provinces and territories grouped into a single row.

We can do this with a function in {dplyr}, `case_when()`.

```
canpop <- read_csv(dpjr::dpjr_data("canpop.csv"))
canpop
```

```
## # A tibble: 13 x 2
##    province_territory        population
##    <chr>                          <dbl>
##  1 Ontario                     13448494
##  2 Quebec                       8164361
##  3 British Columbia             4648055
##  4 Alberta                      4067175
##  5 Manitoba                     1278365
##  6 Saskatchewan                 1098352
##  7 Nova Scotia                   923598
##  8 New Brunswick                 747101
##  9 Newfoundland and Labrador     519716
## 10 Prince Edward Island          142907
## 11 Northwest Territories          41786
## 12 Nunavut                        35944
## 13 Yukon                          35874
```

In this first solution, we name the largest provinces separately:

```
canpop |>
  mutate(pt_grp = case_when(
    # evaluation ~ new value
    province_territory == "Ontario" ~ "Ontario",
    province_territory == "Quebec" ~"Quebec",
    province_territory == "British Columbia" ~ "British Columbia",
    province_territory == "Alberta" ~ "Alberta",
    # all others get recoded as "other"
    TRUE ~ "other"
  ))
```

```
## # A tibble: 13 x 3
##    province_territory        population pt_grp
##    <chr>                          <dbl> <chr>
##  1 Ontario                     13448494 Ontario
##  2 Quebec                       8164361 Quebec
##  3 British Columbia             4648055 British Columbia
##  4 Alberta                      4067175 Alberta
##  5 Manitoba                     1278365 other
##  6 Saskatchewan                 1098352 other
##  7 Nova Scotia                   923598 other
##  8 New Brunswick                 747101 other
##  9 Newfoundland and Labrador     519716 other
## 10 Prince Edward Island          142907 other
## 11 Northwest Territories          41786 other
## 12 Nunavut                        35944 other
## 13 Yukon                          35874 other
```

But that's a lot of typing. A more streamlined approach is to put the name of our provinces in a list, and then recode our new variable with the value from the original variable `province_territory`:

```
canpop |>
  mutate(pt_grp = case_when(
    province_territory %in%
      c("Ontario", "Quebec", "British Columbia", "Alberta") ~
      province_territory,
    TRUE ~
      "other"
  ))
```

```
## # A tibble: 13 x 3
##    province_territory        population pt_grp
##    <chr>                          <dbl> <chr>
##  1 Ontario                     13448494 Ontario
##  2 Quebec                       8164361 Quebec
##  3 British Columbia             4648055 British Columbia
##  4 Alberta                      4067175 Alberta
##  5 Manitoba                     1278365 other
##  6 Saskatchewan                 1098352 other
##  7 Nova Scotia                   923598 other
##  8 New Brunswick                 747101 other
##  9 Newfoundland and Labrador     519716 other
## 10 Prince Edward Island          142907 other
## 11 Northwest Territories          41786 other
## 12 Nunavut                        35944 other
## 13 Yukon                          35874 other
```

We could also use a comparison to create a population threshold, in this case four million:

```
canpop <- canpop |>
  mutate(pt_grp = case_when(
    population > 4000000 ~ province_territory,
    TRUE ~ "other"
  ))

canpop
```

```
## # A tibble: 13 x 3
##    province_territory        population pt_grp
##    <chr>                          <dbl> <chr>
##  1 Ontario                     13448494 Ontario
##  2 Quebec                       8164361 Quebec
##  3 British Columbia             4648055 British Columbia
##  4 Alberta                      4067175 Alberta
##  5 Manitoba                     1278365 other
##  6 Saskatchewan                 1098352 other
##  7 Nova Scotia                   923598 other
##  8 New Brunswick                 747101 other
##  9 Newfoundland and Labrador     519716 other
## 10 Prince Edward Island          142907 other
## 11 Northwest Territories          41786 other
## 12 Nunavut                        35944 other
## 13 Yukon                          35874 other
```

Whatever the approach we choose, that new variable can be used to group a table:

```
canpop |>
  group_by(pt_grp) |>
  summarize(pt_pop = sum(population)) |>
  arrange(desc(pt_pop))
```

```
## # A tibble: 5 x 2
##   pt_grp              pt_pop
##   <chr>                <dbl>
## 1 Ontario           13448494
## 2 Quebec             8164361
## 3 other              4823643
## 4 British Columbia   4648055
## 5 Alberta            4067175
```

12.4.1 Classification systems

As we start to create new variables in our data, it is useful to be aware of various classification systems that already exist. That way, you can:

- save yourself the challenge of coming up with your own classifications for your variables, and
- ensure that your analysis can be compared to other results.

Of course, these classification systems are subject-matter specific.

As one example, in much social and business data, people's ages are often grouped into 5-year "bins", starting with ages 0–4, 5–9, 10–14, and so on. But not all age classification systems follow this; it will be driven by the specific context. The demographics of a workforce will not require a "0–4" category, while an analysis of school-age children might group them by the ages they attend junior, middle, and secondary school. As a consequence, many demographic tables will provide different data sets with the ages grouped in a variety of ways. (For an example, see (Statistics Canada, 2007).)

And be aware that the elements with a classification system can change over time. One familiar example is that the geographical boundaries of cities and countries of the world have changed. City boundaries change, often with the more populous city expanding to absorb smaller communities within the larger legal entity. (The current City of Toronto is an amalgamation of what were six smaller municipalities.) Some regions have a rather turbulent history of changes in their boundaries, and some countries have been created only to disappear a few years later.

If you're working in the area of social and economic research, the national and international statistics agencies provide robust and in-depth classification systems, much of which is explicitly designed to allow for international comparisons. These systems cover a wide range of economic and social statistics. Earlier in this chapter there was an example using the North American Industry Classification System (NAICS), used in Canada, the United States, and Mexico. (Executive Office of the President, Office of Management and Budget, 2022, Statistics Canada (2022); see also Statistics Canada, 2020). A similar classification system exists for Australia and New Zealand (Australian Bureau of Statistics, 2006), and the European Union's Eurostat office publishes a standard classification system for the products that are created through mining, manufacturing, and material recovery (EUROSTAT PRODCOM team, 2022). For international statistics, you can start with the classification system maintained by the United Nations Statistics Division (United Nations Statistics Division, 2020). In Chapter 6, we saw population statistics reported by the administrative geography of England and Wales.

Following these standardized categories in our own work is advantageous. It provides us with a validated structure, and should it be appropriate, gives us access to other data that can be incorporated into our own research.

Your subject area has established classifications; seek them out and use them. Ensuring that your data can be combined with and compared to other datasets will add value to your analysis, and make it easier for future researchers (including future you!) to work with your data. (White et al., 2013)

12.5 Indicator variables

> To include a categorical variable in a regression, a natural approach is to construct an 'indicator variable' for each category. This allows a separate effect for each level of the category, without assuming any ordering or other structure on the categories. (Gelman et al., 2014, p.366)

Indicator variables are also known as dummy variables (Suits, 1957) and "one hot" encoding.

Examples of this approach abound. It can be a useful approach in forecasting methods; Hyndman and Athanasopoulos provide examples where public holidays and days of the week are set as dummy variables (Hyndman and Athanasopoulos, 2021, section 7.4 "Some useful predictors")

It is also a common approach in social data analysis, where categorical variables are used to code information. In the section on "Discrimination and collider bias" in (Cunningham, 2021, pp.106–110), the data are represented in the following way [4]:

```
## # A tibble: 6 x 5
##   female ability discrimination occupation   wage
##    <dbl>   <dbl>          <dbl>      <dbl>  <dbl>
## 1      0  -0.786              0     -0.555  0.162
## 2      1   2.81               1      4.93  11.6
## 3      1  -0.977              1     -4.23  -5.86
## 4      0   1.75               0      3.95  10.4
## 5      0   0.152              0      2.11   4.77
## 6      1  -0.561              1     -1.48  -3.41
```

In this data, the variable `female` is coded as numeric, so it can be used as part of the regression modelling.

More commonly, though, this variable will have been captured and saved in one called `gender` or `sex`, and often as a character string. The source data for the above table might have originally looked like this:

```
tb2 <- tb |>
  dplyr::rename(gender = female) |>
  mutate(gender = case_when(
    gender == 1 ~ "female",
    TRUE ~ "male"
  ))
```

```
tb2
```

```
## # A tibble: 10 x 5
##    gender ability discrimination occupation   wage
```

[4]Note: the data in this table is different from what appears in *Causal Inference: The Mixtape*, since the values in the data in that source are randomly generated.

```
##    <chr>   <dbl>         <dbl>    <dbl> <dbl>
##  1 male   -0.786             0   -0.555  0.162
##  2 female  2.81              1    4.93  11.6
##  3 female -0.977             1   -4.23  -5.86
##  4 male    1.75              0    3.95  10.4
##  5 male    0.152             0    2.11   4.77
##  6 female -0.561             1   -1.48  -3.41
##  7 female  0.272             1   -1.56  -2.47
##  8 male    0.472             0    2.37   4.36
##  9 female  0.627             1    0.238  0.868
## 10 female  0.0128            1   -1.85  -2.54
```

12.5.1 **{fastDummies}**

Because this type of data transformation is common, the package {fastDummies} (Kaplan and Schlegel, 2020) has been created, containing the function `dummy_cols()` (or `dummy_columns()` if you prefer extra typing) that creates indicator variables.

> If you are using the *tidymodels* pipeline, the {recipes} package (Kuhn and Wickham, 2022) contains the `step_dummy()` function, which accomplishes much the same result as `fastDummies::dummy_cols()`. The book *Tidy modelling with R* by Max Kuhn and Julia Silge (Kuhn and Silge, 2022) is recommended; the creation of indicator variables is covered in Chapter 8, "Feature Engineering with recipes".

```
tb2_indicator <- tb2 |>
  fastDummies::dummy_cols()

glimpse(tb2_indicator, width = 65)
```

```
## Rows: 10
## Columns: 7
## $ gender         <chr> "male", "female", "female", "male", "mal~
## $ ability        <dbl> -0.78648, 2.81050, -0.97717, 1.74628, 0.~
## $ discrimination <dbl> 0, 1, 1, 0, 0, 1, 1, 0, 1, 1
```

```
## $ occupation     <dbl> -0.5551, 4.9340, -4.2303, 3.9487, 2.1148~
## $ wage           <dbl> 0.1616, 11.6113, -5.8595, 10.4124, 4.774~
## $ gender_female  <int> 0, 1, 1, 0, 0, 1, 1, 0, 1, 1
## $ gender_male    <int> 1, 0, 0, 1, 1, 0, 0, 1, 0, 0
```

The variable `gender` has now been mutated into two additional variables, `gender_female` and `gender_male`. The names of the new variables are a concatenation of the original variable name and value, separated by an underscore.

Where the value of `gender` is "female", the value of `gender_female` is assigned as "1", and where `gender` is "male", `gender_female` is "0". This parallels what was in the original data. The opposite is true of `gender_male`; the rows where the value of `gender` is "male" `gender_male` has the value "1".

Multicollinearity can occur in a multiple regression model where multiple indicator variables are included, since they are the inverse of one another. Accordingly, an important set of options in the `dummy_cols()` function revolve around removing all but one of the created variables. The argument `remove_first_dummy = TRUE` does just that; in this case, the `gender_female()` variable does not appear in the final result. This is because the new indicator variables are created in alphabetical order ("female" coming before "male"), and the first is dropped.

```
tb2_indicator <- tb2 |>
  fastDummies::dummy_cols(remove_first_dummy = TRUE)

glimpse(tb2_indicator, width = 65)
```

```
## Rows: 10
## Columns: 6
## $ gender         <chr> "male", "female", "female", "male", "mal~
## $ ability        <dbl> -0.78648, 2.81050, -0.97717, 1.74628, 0.~
## $ discrimination <dbl> 0, 1, 1, 0, 0, 1, 1, 0, 1, 1
## $ occupation     <dbl> -0.5551, 4.9340, -4.2303, 3.9487, 2.1148~
## $ wage           <dbl> 0.1616, 11.6113, -5.8595, 10.4124, 4.774~
## $ gender_male    <int> 1, 0, 0, 1, 1, 0, 0, 1, 0, 0
```

Note that the results of this argument can be controlled through the conversion of character variables to factors. To match the original dataframe, we would retain the variable `gender_female`.

```
# create new tb table with "gender" as factor
tb3 <- tb2 |>
```

```
  mutate(gender = as.factor(gender))

levels(tb3$gender)
```

```
## [1] "female" "male"
```

Note that the default order is alphabetical, so "female" remains first. When the argument `remove_first_dummy = TRUE` is applied, we get the same result as before.

```
tb3_1 <- tb3 |>
  fastDummies::dummy_cols(remove_first_dummy = TRUE)

glimpse(tb3_1, width = 65)
```

```
## Rows: 10
## Columns: 6
## $ gender         <fct> male, female, female, male, male, female~
## $ ability        <dbl> -0.78648, 2.81050, -0.97717, 1.74628, 0.~
## $ discrimination <dbl> 0, 1, 1, 0, 0, 1, 1, 0, 1, 1
## $ occupation     <dbl> -0.5551, 4.9340, -4.2303, 3.9487, 2.1148~
## $ wage           <dbl> 0.1616, 11.6113, -5.8595, 10.4124, 4.774~
## $ gender_male    <int> 1, 0, 0, 1, 1, 0, 0, 1, 0, 0
```

By using the `fct_relevel()` function from the {forcats}(Wickham, 2021a) package, we can assign an arbitrary order.

```
# reorder levels

tb3$gender <- fct_relevel(tb3$gender, "male", "female")

levels(tb3$gender)
```

```
## [1] "male"   "female"
```

Rerunning the same code, the `gender_male` variable is dropped.

```
tb3_2 <- tb3 |>
  fastDummies::dummy_cols(remove_first_dummy = TRUE)
```

```
glimpse(tb3_2, width = 65)
```

```
## Rows: 10
## Columns: 6
## $ gender         <fct> male, female, female, male, male, female~
## $ ability        <dbl> -0.78648, 2.81050, -0.97717, 1.74628, 0.~
## $ discrimination <dbl> 0, 1, 1, 0, 0, 1, 1, 0, 1, 1
## $ occupation     <dbl> -0.5551, 4.9340, -4.2303, 3.9487, 2.1148~
## $ wage           <dbl> 0.1616, 11.6113, -5.8595, 10.4124, 4.774~
## $ gender_female  <int> 0, 1, 1, 0, 0, 1, 1, 0, 1, 1
```

There is also an option in `fastDummies::dummy_cols()` to `remove_most_frequent_dummy = TRUE`. In our discrimination data, there are more observations coded as "female", so the variable `gender_female` is not included.

```
tb2_frequent <- tb2 |>
  fastDummies::dummy_cols(remove_most_frequent_dummy = TRUE)

glimpse(tb2_frequent, width = 65)
```

```
## Rows: 10
## Columns: 6
## $ gender         <chr> "male", "female", "female", "male", "mal~
## $ ability        <dbl> -0.78648, 2.81050, -0.97717, 1.74628, 0.~
## $ discrimination <dbl> 0, 1, 1, 0, 0, 1, 1, 0, 1, 1
## $ occupation     <dbl> -0.5551, 4.9340, -4.2303, 3.9487, 2.1148~
## $ wage           <dbl> 0.1616, 11.6113, -5.8595, 10.4124, 4.774~
## $ gender_male    <int> 1, 0, 0, 1, 1, 0, 0, 1, 0, 0
```

Another argument in the function is `remove_selected_columns`. The default value is `FALSE`, but if the argument is set to `TRUE`, the source column is excluded from the output.

```
tb2_select <- tb2 |>
  fastDummies::dummy_cols(remove_selected_columns = TRUE)

glimpse(tb2_select, width = 65)
```

```
## Rows: 10
```

```
## Columns: 6
## $ ability        <dbl> -0.78648, 2.81050, -0.97717, 1.74628, 0.~
## $ discrimination <dbl> 0, 1, 1, 0, 0, 1, 1, 0, 1, 1
## $ occupation     <dbl> -0.5551, 4.9340, -4.2303, 3.9487, 2.1148~
## $ wage           <dbl> 0.1616, 11.6113, -5.8595, 10.4124, 4.774~
## $ gender_female  <int> 0, 1, 1, 0, 0, 1, 1, 0, 1, 1
## $ gender_male    <int> 1, 0, 0, 1, 1, 0, 0, 1, 0, 0
```

Some important things to note about `dummy_cols()`

- the function will create as many indicator variables as there are values in the character or factor variable
- the function will create indicator variables for all of the character and factor variables in the data, unless otherwise specified.

Let's look at how `dummy_cols()` behaves with two of the variables in {palmer-penguins}:

```
library(palmerpenguins)

set.seed(1729) # Ramanujan's taxicab number: 1^3 + 12^3 = 9^3 + 10^3

penguin_subset <- penguins |>
  slice_sample(n = 6) |>
  dplyr::select(species, island) |>
  arrange(species)

penguin_subset
```

```
## # A tibble: 6 x 2
##   species   island
##   <fct>     <fct>
## 1 Adelie    Biscoe
## 2 Adelie    Torgersen
## 3 Chinstrap Dream
## 4 Gentoo    Biscoe
## 5 Gentoo    Biscoe
## 6 Gentoo    Biscoe
```

```
penguin_subset_1 <- penguin_subset |>
  fastDummies::dummy_cols()

glimpse(penguin_subset_1, width = 65)
```

```
## Rows: 6
## Columns: 8
## $ species            <fct> Adelie, Adelie, Chinstrap, Gentoo, Ge~
## $ island             <fct> Biscoe, Torgersen, Dream, Biscoe, Bis~
## $ species_Adelie     <int> 1, 1, 0, 0, 0, 0
## $ species_Chinstrap  <int> 0, 0, 1, 0, 0, 0
## $ species_Gentoo     <int> 0, 0, 0, 1, 1, 1
## $ island_Biscoe      <int> 1, 0, 0, 1, 1, 1
## $ island_Dream       <int> 0, 0, 1, 0, 0, 0
## $ island_Torgersen   <int> 0, 1, 0, 0, 0, 0
```

First, you will notice that we started with two factor variables, `species` and `island`; we now have indicator variables for both.

And because there are three species of penguins and three islands, we have three indicator variables for both of the original variables.

In the instance where we have multiple character or factor variables in our source data, we can control which variables get the `dummy_cols()` treatment with the option, `select_columns`. In the example below, the result provides us with an indicator variable for `species` but not `island`.

In addition, rather than dropping the first created variable, we can specify that the most frequently occurring value be removed with the argument `remove_most_frequent_dummy = TRUE`. Because there are more Gentoo penguins in our sample of the data, the indicator variable `species_Gentoo` is not created.

```
penguin_subset_2 <- penguin_subset |>
  fastDummies::dummy_cols(
    select_columns = "species",
    remove_most_frequent_dummy = TRUE)

glimpse(penguin_subset_2, width = 65)
```

```
## Rows: 6
## Columns: 4
## $ species            <fct> Adelie, Adelie, Chinstrap, Gentoo, Ge~
## $ island             <fct> Biscoe, Torgersen, Dream, Biscoe, Bis~
```

```
## $ species_Adelie    <int> 1, 1, 0, 0, 0, 0
## $ species_Chinstrap <int> 0, 0, 1, 0, 0, 0
```

12.6 Missing values

Frequently, our data will have missing values. Some of this will be gaps in the data, and in other cases it will be due to the structure of the source data file.

There are a variety of solutions to dealing with gaps in data values (sometimes called "missingness"), from listwise deletion (removing all rows with missing values) to complex algorithms that apply methods to impute the best estimate of the missing value. There is in-depth literature on this topic (see (Little, 2020), (Gelman et al., 2014, chapter 18), and (Gelman and Hill, 2007, chapter 25)); what is presented here are some solutions to common and relatively simple problems.

12.6.1 Missing as formatting

A common spreadsheet formatting practice that makes the tables human-readable but less analysis-ready is to have a sub-section heading, often in a separate column. In the example Excel file that we will use for this, the sheet "Report" contains data on tuition fees paid by international students in British Columbia, arranged in a wide format. The name of the individual institution is in the second column (in Excel nomenclature, "B"), and the region ("Economic Development Region") is in the first column. However, the name of the region only appears in the row above the institutions in that region. This makes for a nicely formatted table, but isn't very helpful if we are trying to calculate average tuition fees by region[5].

Our first step is to read in the data. Note that the code uses the Excel naming convention to define the rectangular range to be read and omits the title rows at the top and the notes at the bottom of the sheet.

```
tuition_data <-
  readxl::read_excel(dpjr::dpjr_data("intl_tuition_fees_bc.xlsx"),
                     sheet = "Report",
```

[5]The creators of the file have provided us with a tidy sheet named "Data" and deserve kudos for sharing the data in a variety of layouts. But please be aware that this approach is the exception, not the rule.

```
                    range = "A5:L36")

head(tuition_data)
```

```
## # A tibble: 6 x 12
##   Economic Development~1 Institution `AY 2012/13` `AY 2013/14`
##   `AY 2014/15`
##   <chr>                  <chr>              <dbl>        <dbl>        <dbl>
## 1 Mainland/Southwest     <NA>                  NA           NA           NA
## 2 <NA>                  British Co~        17611.       17963.       18323.
## 3 <NA>                   Capilano U~        15000        15750        16170
## 4 <NA>                   Douglas Co~        14400        15000        15300
## 5 <NA>                  Emily Carr~        15000.       15600        16224.
## 6 <NA>                  Justice In~        13592.       13864.       14141.
## # i abbreviated name: 1: `Economic Development Region`
## # i 7 more variables: `AY 2015/16` <dbl>, `AY 2016/17` <dbl>,
## #   `AY 2017/18` <dbl>, `AY 2018/19` <dbl>, `AY 2019/20` <dbl>,
## #   `AY 2020/21` <dbl>, `AY 2021/22` <dbl>
```

The contents of the file have been read correctly, but there are many "NA" values in the `Economic Development Region` variable. This is due to the manner in which the data was structured in the source Excel file.

The `fill()` function, in the {tidyr} package (Wickham, 2021b), provides the solution to this. The function replaces a missing value with either the previous or the next value in a variable. In the example here, it will replace all of the "NA" values with the region name above, stopping when it comes to the next non-NA record.

```
tuition_data_filled <- tuition_data |>
  fill("Economic Development Region")

head(tuition_data_filled)
```

```
## # A tibble: 6 x 12
##   Economic Development~1 Institution `AY 2012/13` `AY 2013/14`
##   `AY 2014/15`
##   <chr>                  <chr>              <dbl>        <dbl>        <dbl>
## 1 Mainland/Southwest     <NA>                  NA           NA           NA
## 2 Mainland/Southwest    British Co~        17611.       17963.       18323.
## 3 Mainland/Southwest    Capilano U~        15000        15750        16170
## 4 Mainland/Southwest    Douglas Co~        14400        15000        15300
## 5 Mainland/Southwest    Emily Carr~        15000.       15600        16224.
```

```
## 6 Mainland/Southwest    Justice In~     13592.    13864.   14141.
## # i abbreviated name: 1: `Economic Development Region`
## # i 7 more variables: `AY 2015/16` <dbl>, `AY 2016/17` <dbl>,
## #   `AY 2017/18` <dbl>, `AY 2018/19` <dbl>, `AY 2019/20` <dbl>,
## #   `AY 2020/21` <dbl>, `AY 2021/22` <dbl>
```

The next step is to use the {tidyr} function `drop_na()` to omit any row that contains an NA value. In this case, we want to drop what were the header rows that contained just the region name in the first column and no values in the other columns.

These two solutions yield the same result:

```
drop_na(tuition_data_filled)
```

```
tuition_data_filled |>
  drop_na()
```

```
## # A tibble: 25 x 12
##   `Economic Development Region` Institution    `AY 2012/13` `AY 2013/14`
##   <chr>                         <chr>                 <dbl>        <dbl>
## 1 Mainland/Southwest            British Columbi~     17611.       17963.
## 2 Mainland/Southwest            Capilano Univer~     15000        15750
## 3 Mainland/Southwest            Douglas College      14400        15000
## 4 Mainland/Southwest            Emily Carr Univ~     15000.       15600
## 5 Mainland/Southwest            Justice Institu~     13592.       13864.
## 6 Mainland/Southwest            Kwantlen Polyte~     15000        15750
## 7 Mainland/Southwest            Langara College      16500        16950
## 8 Mainland/Southwest            Simon Fraser Un~     16236        17862
## 9 Mainland/Southwest            University of B~     22622.       23300.
## 10 Mainland/Southwest           University of t~     12720        13350
## # i 15 more rows
## # i 8 more variables: `AY 2014/15` <dbl>, `AY 2015/16` <dbl>,
## #   `AY 2016/17` <dbl>, `AY 2017/18` <dbl>, `AY 2018/19` <dbl>,
## #   `AY 2019/20` <dbl>, `AY 2020/21` <dbl>, `AY 2021/22` <dbl>
```

It's worth noting that the `drop_na()` function provides an argument where we can specify columns. This gives us the flexibility to drop some rows where there is an "NA" in a specific variable, while keeping other rows that have a value in that variable but not in other variables.

12.6.2 Missing numeric values

"Missing data are unobserved values that would be meaningful for analysis if observed; in other words, a missing value hides a meaningful value." (Little, 2020)

The functions used above for filling missing text values can also be used for missing numeric values. But when dealing with the replacement of numeric values, we need to be mindful that the values we use will be incorporated into models and analysis; we need to be clear that the filled values are *estimates*, including sufficient documentation in our methodology section.

Furthermore, there is a great deal of literature on the topic of imputation of missing values. As researchers, we need to be aware of *why* the values are missing; are we dealing with values that are missing "at completely random," "at random," or "not at random?" Delving into the literature, you'll discover that these terms are precisely defined, and the solutions we adopt to deal with missingness will depend on the reason (or reasons) why the data are missing. (Little, 2020)

For the examples presented here, we will assume that you have already given significant thought to these methodological questions. As we discussed earlier in this chapter, *how* we clean data is influenced by our research questions. As well, we need to consider our cleaning strategy in light of why the raw data are the way they are.

With that preamble, let's move to working with some real data. In this example, the data series is the number of salaried employees in the Accommodation services industry in British Columbia, Canada, for the years 2001 through 2020. The data are collected and published by Statistics Canada, and for some years the data are judged to be "too unreliable to be published." This leaves some gaps in our data.

```
## # A tibble: 6 x 4
##   label    ref_year REF_DATE   VALUE
##   <chr>       <dbl> <chr>      <dbl>
## 1 Salaried     2001 2001-01-01  4458
## 2 Salaried     2002 2002-01-01  4392
## 3 Salaried     2003 2003-01-01  4124
## 4 Salaried     2004 2004-01-01  4828
## 5 Salaried     2005 2005-01-01  4069
## 6 Salaried     2006 2006-01-01  5048
```

We will employ Exploratory Data Analysis, and plot the data:

```
ggplot(naics721_sal, aes(x = ref_year, y = VALUE)) +
  geom_line()
```

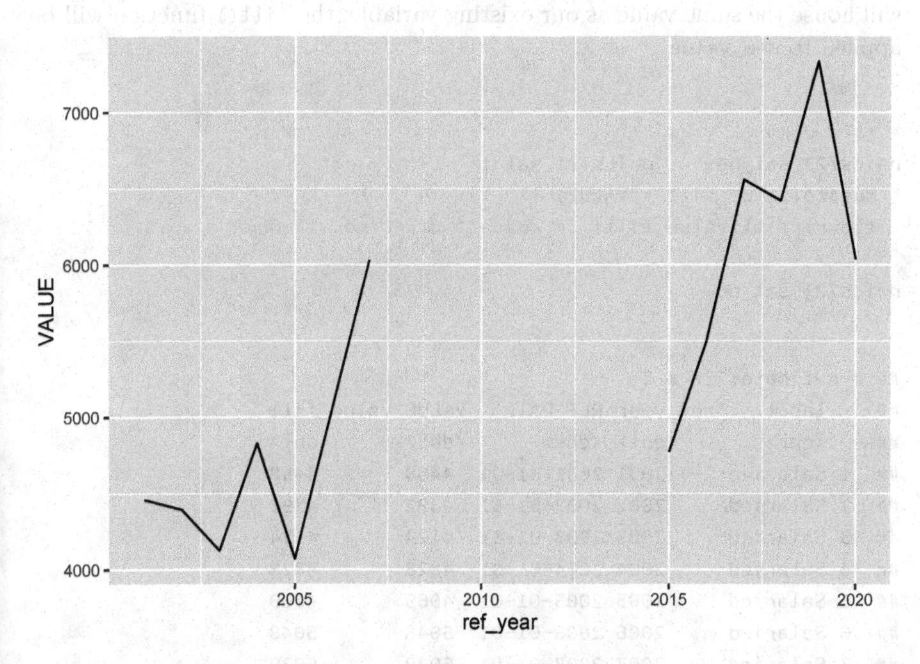

Our first approach will be to simply drop those years that are missing a value. This might help with some calculations (for example, putting a `na.rm = TRUE` argument in a `mean()` function is doing just that), but it will create a misleading plot—there will be the same distance between the 2009 and 2015 points as there is between 2015 and 2016.

```
naics721_sal |>
  tidyr::drop_na() |>
  # show only first five rows
  slice_head(n = 5)
```

```
## # A tibble: 5 x 4
##   label    ref_year REF_DATE   VALUE
##   <chr>       <dbl> <chr>      <dbl>
## 1 Salaried     2001 2001-01-01  4458
## 2 Salaried     2002 2002-01-01  4392
## 3 Salaried     2003 2003-01-01  4124
## 4 Salaried     2004 2004-01-01  4828
## 5 Salaried     2005 2005-01-01  4069
```

Alternatively, we can fill the series with the previous valid entry.

Similar to what we saw above with text, the same `fill()` function can be applied to numeric variables. We will first create a variable `new_value` that will house the same value as our existing variable; the `fill()` function will be applied to `new_value`.

```
# fill down
naics721_sal_new <- naics721_sal |>
  mutate(value_fill = VALUE) |>
  tidyr::fill(value_fill)  # default direction is down

naics721_sal_new
```

```
## # A tibble: 20 x 5
##    label    ref_year REF_DATE   VALUE value_fill
##    <chr>       <dbl> <chr>      <dbl>      <dbl>
##  1 Salaried     2001 2001-01-01  4458       4458
##  2 Salaried     2002 2002-01-01  4392       4392
##  3 Salaried     2003 2003-01-01  4124       4124
##  4 Salaried     2004 2004-01-01  4828       4828
##  5 Salaried     2005 2005-01-01  4069       4069
##  6 Salaried     2006 2006-01-01  5048       5048
##  7 Salaried     2007 2007-01-01  6029       6029
##  8 Salaried     2008 2008-01-01    NA       6029
##  9 Salaried     2009 2009-01-01  5510       5510
## 10 Salaried     2010 2010-01-01    NA       5510
## 11 Salaried     2011 2011-01-01    NA       5510
## 12 Salaried     2012 2012-01-01    NA       5510
## 13 Salaried     2013 2013-01-01    NA       5510
## 14 Salaried     2014 2014-01-01    NA       5510
## 15 Salaried     2015 2015-01-01  4772       4772
## 16 Salaried     2016 2016-01-01  5492       5492
## 17 Salaried     2017 2017-01-01  6557       6557
## 18 Salaried     2018 2018-01-01  6420       6420
## 19 Salaried     2019 2019-01-01  7334       7334
## 20 Salaried     2020 2020-01-01  6036       6036
```

Now when we plot the data, the missing points are shown:

```
ggplot(naics721_sal_new, aes(x = ref_year, y = value_fill)) +
  geom_line()
```

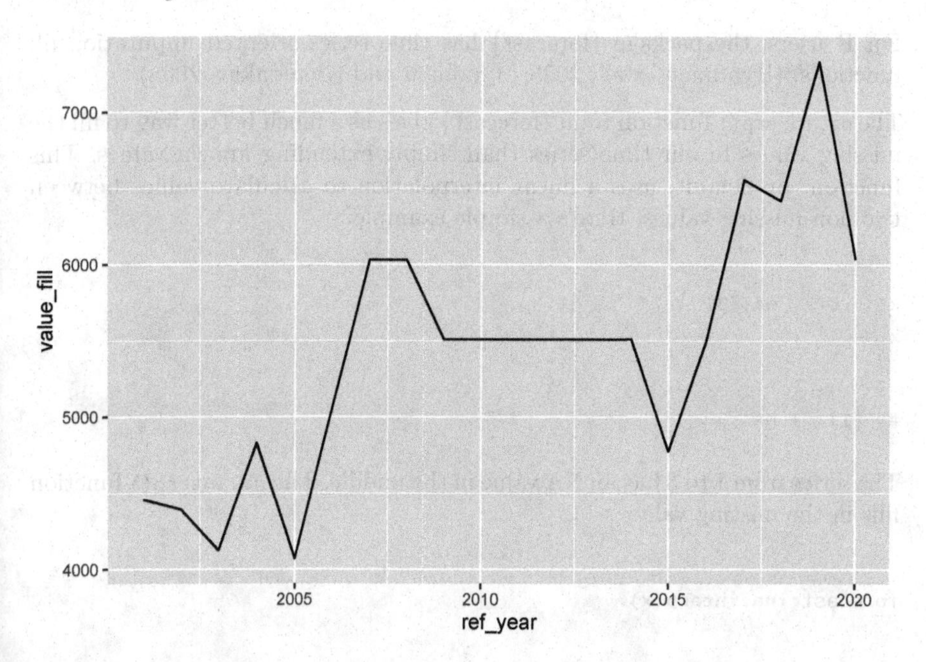

There are a few different options to determine which way to fill; the default is `.direction = "down"`.

```
# specified
# down
naics721_sal |>
  mutate(value_down = VALUE) |>
  tidyr::fill(value_down, .direction = "down")  # direction is specified
     as down
# up
naics721_sal |>
  mutate(value_up = VALUE) |>
  tidyr::fill(value_up, .direction = "up")  # direction is specified as
  up
```

12.6.3 Time series imputation

You may already have realized that the values we filled above are drawn from a time series. Economists, biologists, and many other researchers who work with time series data have built up a large arsenal of tools to work with time series data, including robust methodologies for dealing with missing values.

For R users, the package {forecast} has time series-oriented imputation fill functions. (Hyndman et al., 2023; Hyndman and Khandakar, 2008)

The `na.interp()` function from {forecast} gives us a much better way to fill the missing values in our time series than simply extending known values. This function, by default, uses a linear interpolation to calculate values between two non-missing values. Here's a simple example:

```
x <- c(1, NA, 3)
x
```

```
## [1]  1 NA  3
```

The series from 1 to 3 has an NA value in the middle. The `na.interp()` function fills in the missing value.

```
forecast::na.interp(x)
```

```
## Registered S3 method overwritten by 'quantmod':
##   method            from
##   as.zoo.data.frame zoo

## Time Series:
## Start = 1
## End = 3
## Frequency = 1
## [1] 1 2 3
```

For longer series, a similar process occurs:

```
y <- c(1, NA, NA, 3)
y
```

```
## [1]  1 NA NA  3
```

```
forecast::na.interp(y)
```

```
## Time Series:
## Start = 1
```

```
## End = 4
## Frequency = 1
## [1] 1.000 1.667 2.333 3.000
```

You will notice that the function has calculated replacement values that give us a linear series.

For our count of salaried employees, here's what we get when we create a new value, `value_new`, with this function:

```
naics721_sal <- naics721_sal |>
  mutate(value_interp = forecast::na.interp(VALUE))

naics721_sal
```

```
## # A tibble: 20 x 5
##    label    ref_year REF_DATE   VALUE value_interp
##    <chr>       <dbl> <chr>      <dbl>        <dbl>
##  1 Salaried     2001 2001-01-01  4458         4458
##  2 Salaried     2002 2002-01-01  4392         4392
##  3 Salaried     2003 2003-01-01  4124         4124
##  4 Salaried     2004 2004-01-01  4828         4828
##  5 Salaried     2005 2005-01-01  4069         4069
##  6 Salaried     2006 2006-01-01  5048         5048
##  7 Salaried     2007 2007-01-01  6029         6029
##  8 Salaried     2008 2008-01-01    NA         5770.
##  9 Salaried     2009 2009-01-01  5510         5510
## 10 Salaried     2010 2010-01-01    NA         5387
## 11 Salaried     2011 2011-01-01    NA         5264
## 12 Salaried     2012 2012-01-01    NA         5141
## 13 Salaried     2013 2013-01-01    NA         5018
## 14 Salaried     2014 2014-01-01    NA         4895
## 15 Salaried     2015 2015-01-01  4772         4772
## 16 Salaried     2016 2016-01-01  5492         5492
## 17 Salaried     2017 2017-01-01  6557         6557
## 18 Salaried     2018 2018-01-01  6420         6420
## 19 Salaried     2019 2019-01-01  7334         7334
## 20 Salaried     2020 2020-01-01  6036         6036
```

Plotting the data with the interpolated values, there are no gaps.

```
ggplot(naics721_sal, aes(x = ref_year, y = value_interp)) +
  geom_line()
```

```
## Don't know how to automatically pick scale for object of type <ts>.
## Defaulting to continuous.
```

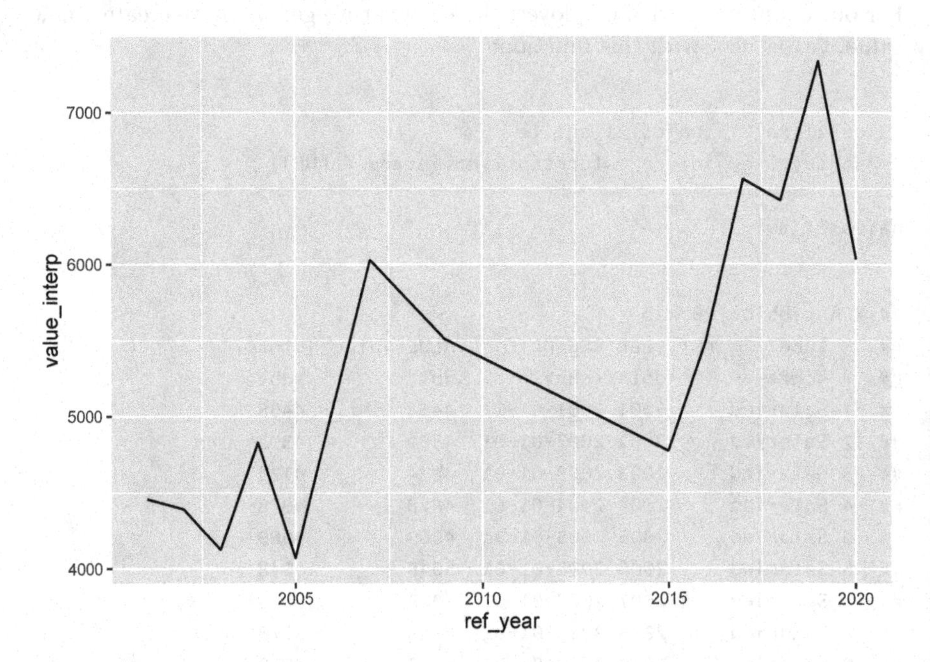

The `na.interp()` function also has the capacity to create non-linear models.

Other examples of time series missing value imputation functions can be found in the {imputeTS} package (Moritz and Bartz-Beielstein, 2017), which offers multiple imputation algorithms and plotting functions. A fantastic resource for using R for time series analysis, including a section on imputation of missing values, is (Hyndman and Athanasopoulos, 2021).

12.7 Creating labelled factors from numeric variables

In Chapter 7 we imported statistical software packages such as SPSS, SAS, or Stata and saw the benefits, in certain circumstances, of having labelled variables. These are categorical variables, where there are a limited number of categories.

In some cases, there might be one or more values assigned to categories we might want to designate as "missing"—a common example can be found in surveys, where respondents are given the option of answering "don't know" and "not applicable." While these responses might be interesting in and of themselves, in some contexts we will want to assign them as missing.

In other cases, numeric values are used to store the categorical responses.

The Joint Canada/United States Survey of Health (JCUSH) file we saw in Chapter 5 importing fixed width files gives us a good example of where categories are stored as numerical variables. For example, the variable `SPJ1_TYP` is either a "1" for responses from the Canadian sample and a "2" for the sample from the United States.

We will use this file to create labelled factors from the numeric values in the variables we imported. Here's the code we used previously to read the file:

```
jcush <- readr::read_fwf(dpjr::dpjr_data("JCUSH.txt"),
    fwf_cols(
      SAMPLEID = c(1, 12),
      SPJ1_TYP = c(13, 13),
      GHJ1DHDI = c(32, 32),
      SDJ1GHED = c(502, 502)
      ),
    col_types = list(
      SAMPLEID = col_character()
    ))

head(jcush)
```

```
## # A tibble: 6 x 4
##   SAMPLEID     SPJ1_TYP GHJ1DHDI SDJ1GHED
##   <chr>           <dbl>    <dbl>    <dbl>
## 1 100003330835        1        3        4
## 2 100004903392        1        3        2
## 3 100010137168        1        2        1
## 4 100010225523        1        3        3
## 5 100011623697        1        2        2
## 6 100013652729        1        4        3
```

The values for these variables are as follows:

Name	Variable	Code	Value
SAMPLEID	Household identifier	unique number	
SPJ1_TYP	Sample type [country]	1	Canada
		2	United States
GHJ1DHDI	Health Description Index	0	Poor
		1	Fair
		2	Good
		3	Very Good
		4	Excellent
		9	Not Stated
SDJ1GHED	Highest level of post-secondary education attained	1	Less than high school
		2	High school degree or equivalent (GED)
		3	Trades Cert, Voc. Sch./Comm.Col./CEGEP
		4	Univ or Coll. Cert. incl. below Bach.
		9	Not Stated

The variable `GHJ1DHDI` has the respondent's self-assessment of their overall health. Here's a summary table of the responses:

```
jcush |>
  group_by(GHJ1DHDI) |>
  tally()
```

```
## # A tibble: 6 x 2
##   GHJ1DHDI     n
##      <dbl> <int>
## 1        0   391
## 2        1   891
## 3        2  2398
## 4        3  2914
## 5        4  2082
## 6        9    12
```

But it would be much more helpful and efficient if we attach the value labels to the dataframe.

One strategy is to create a new factor variable, that uses the value description instead of the numeric representation. For this, we will use the `fct_recode()` function from the {forcats} package(Wickham, 2021a).

```
library(forcats)

jcush_forcats <- jcush |>
  # 1st mutate a new factor variable with the original values
  mutate(health_desc_fct = as.factor(GHJ1DHDI)) |>
  # 2nd recode the values
  mutate(
    health_desc_fct = fct_recode(
      health_desc_fct,
      "Poor" = "0",
      "Fair" = "1",
      "Good" = "2",
      "Very Good" = "3",
      "Excellent" = "4",
      NULL = "9"
    )
  )
```

The `levels()` function allows us to inspect the labels. Note that because we specified the original value of "9" as `NULL`, it does not appear in the list.

```
levels(jcush_forcats$health_desc_fct)
```

```
## [1] "Poor"      "Fair"      "Good"      "Very Good" "Excellent"
```

When we tally the results by our new variable, the `GHJ1DHDI` records with the value 9 are shown as "NA".

```
jcush_forcats |>
  group_by(GHJ1DHDI, health_desc_fct) |>
  tally()
```

```
## # A tibble: 6 x 3
## # Groups:   GHJ1DHDI [6]
##   GHJ1DHDI health_desc_fct     n
##      <dbl> <fct>           <int>
## 1        0 Poor              391
## 2        1 Fair              891
## 3        2 Good             2398
## 4        3 Very Good        2914
## 5        4 Excellent        2082
## 6        9 <NA>               12
```

Another option is provided in the {haven} package (Wickham and Miller, 2021).

```
library(haven)

jcush_haven <- jcush |>
  mutate(health_desc_lab = labelled(
    GHJ1DHDI,
    c(
      "Poor" = 0,
      "Fair" = 1,
      "Good" = 2,
      "Very Good" = 3,
      "Excellent" = 4,
      "Not Stated" = 9
    )
  ))
```

Note that our new variable is a "S3: haven_labelled" class, and the `ls.str()` function shows it as "dbl+lbl":

```
ls.str(jcush_haven)
```

```
## GHJ1DHDI :  num [1:8688] 3 3 2 3 2 4 2 4 0 2 ...
## health_desc_lab :  dbl+lbl [1:8688] 3, 3, 2, 3, 2, 4, 2, 4, 0, 2, 4, 4, 4, 2, 2, 2, 1, 4...
## SAMPLEID :  chr [1:8688] "100003330835" "100004903392" "100010137168" ...
## SDJ1GHED :  num [1:8688] 4 2 1 3 2 3 2 1 3 2 ...
## SPJ1_TYP :  num [1:8688] 1 1 1 1 1 1 1 1 1 1 ...
```

FIGURE 12.5 *ls.str(jcush_haven).*

The `levels()` function can be used to inspect the result of this mutate.

```
head(jcush_haven$health_desc_lab)
```

```
## Found more than one class "haven_labelled" in cache; using the first,
##  from namespace 'memisc'

## Also defined by 'haven'

## Found more than one class "haven_labelled" in cache; using the first,
## from namespace 'memisc'

## Also defined by 'haven'

## <labelled<double>[6]>
## [1] 3 3 2 3 2 4
##
## Labels:
##  value      label
##      0       Poor
##      1       Fair
##      2       Good
##      3  Very Good
##      4  Excellent
##      9 Not Stated
```

A third option is to use the functions within the {labelled} package (Larmarange, 2022).

```
jcush_labelled <- jcush |>
  mutate(health_desc_lab = labelled(
    GHJ1DHDI,
    c(
      "Poor" = 0,
      "Fair" = 1,
      "Good" = 2,
      "Very Good" = 3,
      "Excellent" = 4,
      "Not Stated" = 9
    )
  ))
```

To view a single variable, use:

```
head(jcush_labelled$health_desc_lab)
```

```
## <labelled<double>[6]>
## [1] 3 3 2 3 2 4
##
## Labels:
##  value      label
##      0       Poor
##      1       Fair
##      2       Good
##      3  Very Good
##      4  Excellent
##      9 Not Stated
```

This package also allows for a descriptive name of the variable to be appended with the `var_lab()` function.

13

Recap

13.1 Some key points

In this book, we have worked through numerous examples of preparing data for analysis, whether that's summary tabulations or modelling methods such as linear regression with indicator variables or time series analysis and forecasting.

Some key points worth re-iterating:

- The data you get is not going to be the data you need. It is going to take work—thoughtful work—to prepare it for analysis.
- The process of preparing data is not a linear, step-by-step process. It involves iterating and looping back to previous steps to address issues with the data. Iteration begins with the *import* step and continues through *validation* and *cleaning.*
- Every one of the elements in the preparation process requires you, the researcher, to make decisions based on your judgement about how best to fulfill the research objective. Starting with defining which variables to import through how to resolve missing values in the data, you have to make informed and thoughtful decisions based on how the data is going to be used.
- Get as far as you can, as quickly as you can, with the import function. The fewer post-import changes you need to make, the better.
- Getting the data into a tidy structure is essential.
- Documentation is essential, both in the code and in the project folder.
  - State the *why* of the code, not the *what.*
  - Create a README file with the project objective and other key pieces of information.
  - Create a data dictionary.

  - The audience for this documentation might be a collaborator you already know or a future user of the data whom you have not met. The audience almost certainly includes you in the future—next year, next month, or tomorrow.

- The data you receive will come in a variety of formats, each format having its own characteristics. Being familiar with what various formats bring (or don't bring) to the analysis can help you be more efficient. For example, labelled formats such as SPSS and SAS files carry with them valuable meta data about the variables and values; taking advantage of that additional information can create greater insights in the analysis.
- Clean data is complete, consistent, and accurate. With these three qualities it is *believable.* Validation and cleaning will strive to improve the data in all three of these dimensions.
- Exploring and validating data require subject matter knowledge.
- Use a combination of approaches, including visualization and structured testing, to first identify the ways in which your data is dirty, and then after taking the necessary steps, to ensure that the data is clean.
- The steps necessary to clean the data are going to be contingent on the research question. The same source data may need to be cleaned in completely different ways for two different research questions.
- In some instances, replacement of missing values helps resolve structural problems with our data. In other cases, replacing missing values can introduce changes to the results of our analysis. Be careful as you introduce replacement into your data cleaning.

13.2 Where to from here?

We started this book with the metaphor of the lighthouse, which guides ships in the right direction and keeps them from running aground. I hope that the principles and the examples in this book achieve the same, in the context of preparing data for analysis.

The examples in this book provide a guide to some of the types of challenges that you might confront when preparing data for analysis. But as you embark on your own data preparation journey, you will run into problems that are different than the ones here. The chances are high that the examples here won't be enough to solve the problem. You may need to think through the problem in a different way, in order to come to the solution. You may require other tools, whether other R packages or a different programming language. You might encounter different data storage methods than what has been introduced here; this is particularly likely if you start working with very large databases.

Our knowledge is always incomplete. This is why we need to keep learning and practicing our craft. If you don't find data preparation challenges in your job or the courses you take, you may want to adapt the examples in this book to other data. This might be finding an Excel file that uses colour as a variable. Or perhaps there's a SQL database that requires you to build a relation between two tables, but those two don't share a common variable, so you have to connect both to a third table that can then link to both.

The growth of the R ecosystem, including the tidyverse, has been astounding. There is a high likelihood that the problems you encounter will have already been seen by someone else. And that someone may have written a blog post about their solution or written a package with a generalized solution.

I hope that the examples in this book give you the confidence to venture forward and to face those challenges as they appear.

Bibliography

Abeysooriya, M., Soria, M., Kasu, M. S., and Ziemann, M. (2021). Gene name errors: Lessons not learned. *PLOS Computational Biology*, 17(7):1–13.

Abeysundera, M. (2015). Using total survey error to study mode effect and other applications. In *Proceedings of the Statistical Society of Canada Annual Meeting, Survey Methods Section*.

Anscombe, F. J. (1973). Graphs in statistical analysis. *The American Statistician*, 27(1):17–21.

Ashton, D. (2018). Where's my T-Shirt? Supply chain forecasting in fashion. Accessed: 2020-07-09.

Au, R. (2020a). Data cleaning is analysis, not grunt work. Accessed: 2020-09-15.

Au, R. (2020b). Let's get intentional about documentation. Accessed: 2020-09-29.

Australian Bureau of Statistics (2006). Australian and New Zealand Standard Industrial Classification (ANZSIC). Technical report, Australian Bureau of Statistics. Accessed: 2023-06-12.

Boysel, S. and Vaughan, D. (2023). *fredr: An R Client for the 'FRED' API*. R package version 2.1.0.9000.

British Columbia Ferry Services Inc. (2021). Annual report to the british columbia ferries commissioner. Technical report, British Columbia Ferry Services Inc. Accessed: 2024-02-17.

Broman, K. and Woo, K. (2017). Data organization in spreadsheets. *The American Statistician*, 72(1):2–10.

Bryan, J. (2016a). sanesheets.

Bryan, J. (2016b). spreadsheets.

Bryan, J. (2017). *gapminder: Data from Gapminder*. R package version 0.3.0.

Bryan, J. and The STAT 545 TAs (2019). Stat 545.

Colson, E., Coffey, B., Rached, T., and Cruz, L. (2017). Algorithms tour: How data science is woven into the fabric of stitch fix. Accessed: 2020-07-09.

Conway, D. (2010). The data science Venn diagram. Accessed: 2020-07-04.

Cowgill, M., Meers, Z., Lee, J., and Diviny, D. (2023). *readabs: Download and Tidy Time Series Data from the Australian Bureau of Statistics*. R package version 0.4.13.901.

Cunningham, S. (2021). *Causal Inference: The Mixtape*. Yale University Press.

Davenport, T. H. and Harris, J. G. (2007). *Competing on Analytics: The New Science of Winning*. Harvard Business School Press.

Dowle, M., Srinivasan, A., Gorecki, J., Chirico, M., Stetsenko, P., Short, T., Lianoglou, S., Antonyan, E., Bonsch, M., Parsonage, H., Ritchie, S., Ren, K., Tan, X., Saporta, R., Seiskari, O., Dong, X., Lang, M., Iwasaki, W., Wenchel, S., Broman, K., Schmidt, T., Arenburg, D., Smith, E., Cocquemas, F., Gomez, M., Chataignon, P., Blaser, N., Selivanov, D., Riabushenko, A., Lee, C., Groves, D., Possenriede, D., Parages, F., Toth, D., Yaramaz-David, M., Perumal, A., Sams, J., Morgan, M., Quinn, M., javrucebo, marc outins, Storey, R., Saraswat, M., Jacob, M., Schubmehl, M., Vaughan, D., Hocking, T., Silvestri, L., Barrett, T., Hester, J., Damico, A., Freundt, S., Simons, D., de Andrade, E. S., Miller, C., Meldgaard, J. P., Tlapak, V., Ushey, K., Eddelbuettel, D., and Schwen, B. (2023). *data.table: Extension of 'data.frame'*. R package version 1.14.8.

Elders, B. and Oldoni, D. (2020). *tidylog: Logging for 'dplyr' and 'tidyr' Functions*. R package version 1.0.2.

Elections BC (2018). Annual report 2017/18 and service plan 2018/19 - 2020/21. Technical report, Elections BC. Accessed on 2020-08-01.

Elff, M., Lawrence, C. N., Atkins, D., Morgan, J. W., Müller, K., Schoonees, P., and Zeileis, A. (2021). *memisc: Management of Survey Data and Presentation of Analysis Results*. R package version 0.99.30.7.

Ellis, S. E. and Leek, J. T. (2017). How to share data for collaboration. *The American Statistician*, 72(1):53–57.

EUROSTAT PRODCOM team (2022). European business statistics user's manual for PRODCOM. Technical report, Eurostat. Accessed: 2023-06-12.

Executive Office of the President, Office of Management and Budget (2022). North American industry classification system : United States, 2022. Technical report, US Census Bureau. Accessed: 2023-03-07.

Firke, S. (2021). *janitor: Simple Tools for Examining and Cleaning Dirty Data*. R package version 2.1.0.

Friendly, M., Dalzell, C., Monkman, M., Murphy, D., Foot, V., and Zaki-Azat, J. (2020). *Lahman: Sean 'Lahman' Baseball Database*. R package version 8.0-0.

Garmonsway, D. (2023). *unpivotr: Unpivot Complex and Irregular Data Layouts.* R package version 0.6.3.

Garmonsway, D., Wickham, H., Bryan, J., RStudio, and Kalicinski, M. (2022). *tidyxl: ERead Untidy Excel Files.* R package version 1.0.8.

Gelfand, S. and City of Toronto (2022). *opendatatoronto: Access the City of Toronto Open Data Portal.* R package version 0.1.5.

Gelman, A., Carlin, J. B., Stern, H. S., Dunson, D. B., Vehtari, A., and Rubin, D. B. (2014). *Bayesian Data Analysis.* CRC Press, 3rd edition.

Gelman, A. and Hill, J. (2007). *Data Analysis Using Regression and Multilevel/Hierarchical Models.* CRC Press.

Hester, J., Wickham, H., libuv project contributors, Joyent, I., and other Node contributors (2020). *fs: Cross-Platform File System Operations Based on 'libuv'.* R package version 1.5.0.

Holbrook, A. L., Green, M. C., and Krosnick, J. A. (2003). Telephone versus face-to-face interviewing of national probability samples with long questionnaires: Comparisons of respondent satisficing and social desirability response bias. *Public Opinion Quarterly*, 67(1):79–125.

Horst, A. (2020). *palmerpenguins: Palmer Archipelago (Antarctica) Penguin Data.*

Horst, A. M., Hill, A. P., and Gorman, K. B. (2022). Palmer Archipelago Penguins Data in the palmerpenguins R Package - An Alternative to Anderson's Irises. *The R Journal*, 14(1):244–254.

Hudon, C. (2018). Field notes: Building data dictionaries. Accessed: 2023-06-13.

Hyndman, R., Athanasopoulos, G., Bergmeir, C., Caceres, G., Chhay, L., O'Hara-Wild, M., Petropoulos, F., Razbash, S., Wang, E., and Yasmeen, F. (2023). *forecast: Forecasting Functions for Time Series and Linear Models.* R package version 8.20.

Hyndman, R. J. and Athanasopoulos, G. (2021). *Forecasting: Principles and Practice.* OTexts, 3rd edition.

Hyndman, R. J. and Khandakar, Y. (2008). Automatic time series forecasting: the forecast package for R. *Journal of Statistical Software*, 26(3):1–22.

Iannone, R., Cheng, J., Schloerke, B., Hughes, E., Lauer, A., and Seo, J. (2023). *gt: Easily Create Presentation-Ready Display Tables.* https://gt.rstudio.com/, https://github.com/rstudio/gt.

Kaplan, J. and Schlegel, B. (2020). *fastDummies: Fast Creation of Dummy (Binary) Columns and Rows from Categorical Variables.* R package version 1.6.3.

Khreis, H., Johnson, J., Jack, K., Dadashova, B., and Park, E. (2008). Evaluating the Performance of Low-Cost Air Quality Monitors in Dallas, Texas. *Int J Environ Res Public Health*, 19(3):1647.

Knuth, D. E. (1992). *Literate programming.* CSLI lecture notes ; no. 27. Center for the Study of Language and Information, Stanford, Calif.

Kuhn, M. (2020). *modeldata: Data Sets Useful for Modeling Packages.* R package version 0.0.2.

Kuhn, M. and Silge, J. (2022). *Tidy Modeling with R.* O'Reilly Media.

Kuhn, M. and Wickham, H. (2022). *recipes: Preprocessing and Feature Engineering Steps for Modeling.* R package version 1.0.1.

Larmarange, J. (2022). *labelled: Manipulating Labelled Data.* R package version 2.9.1.9000.

Lee, B. D. (2018). Ten simple rules for documenting scientific software. *PLoS Computational Biology*, 14(12).

Little, R. J. A. (2020). *Statistical analysis with missing data.* Wiley series in probability and statistics. Wiley, Hoboken, NJ, 3rd edition.

Marwick, B., Boettiger, C., and Mullen, L. (2018). Packaging data analytical work reproducibly using R (and friends). *The American Statistician*, 72(1):80–88.

Monkman, M. (2019). Same name, different bird. Accessed: 2020-05-19.

Monkman, M. (2024). *dpjr: Companion data for the book The Data Preparation Journey: Finding Your Way With R.* https://github.com/MonkmanMH/dpjr.

Moritz, S. and Bartz-Beielstein, T. (2017). imputeTS: Time Series Missing Value Imputation in R. *The R Journal*, 9(1):207–218.

Murrell, P. (2013). Data intended for human consumption, not machine consumption. In McCallum, Q. E., editor, *Bad Data Handbook*, chapter 3, page 31–51. O'Reilly.

Müller, K. (2022). *DBI: R Database Interface.* R package version 1.1.1.

Müller, K. and Wickham, H. (2021). *tibble: Simple Data Frames.* R package version 3.1.2.

Müller, K., Wickham, H., James, D. A., Falcon, S., Hipp, D. R., Kennedy, D., Mistachkin, J., SQLite Authors, Healy, L., R Consortium, and RStudio (2023). *RSQLite: SQLite Interface for R.* R package version 2.3.1.

Nield, T. (2016). *Getting Started with SQL: A Hands-On Approach for Beginners.* O'Reilly Media.

Ooms, J. (2022). *pdftools: Text Extraction, Rendering and Converting of PDF Documents.* R package version 3.3.0.

Peng, R. D. (2011). Reproducible research in computational science. *Science*, 334:1226–1227. 2011-12-02.

Perez, C. C. (2019). *Invisible Women: Data Bias in a World Designed for Men.* Abrams Press.

Polonsky, J. A., Baidjoe, A., Kamvar, Z. N., Cori, A., Durski, K., Edmunds, J. W., Eggo, R. M., Funk, S., Kaiser, L., Keating, P., le Polain de Waroux, O., Marks, M., Moraga, P., Morgan, O., Nouvellet, P., Ratnayake, R., Roberts, C. H., Whitworth, J., and Jombart, T. (2019). Outbreak analytics: a developing data science for informing the response to emerging pathogens. *Philosophical Transactions of the Royal Society B*, 374:20180276. 2019-05-20.

R Core Team (2021). *R: A Language and Environment for Statistical Computing.* R Foundation for Statistical Computing, Vienna, Austria.

Research Data Management Service Group (2021). Guide to writing "readme" style metadata. Technical report, Cornell University. Accessed: 2021-10-12.

Smith, D. (2017). Reproducible data science with R. Accessed: 2020-05-18.

St-Pierre', M. and Béland, Y. (2004). Mode effects in the Canadian Community Health Survey: A comparison of CAPI and CATI. In *Proceedings of the American Statistical Association Meeting, Survey Research Methods.*

Statistics Canada (2007). Age of person. Technical report, Statistics Canada. Accessed: 2020-08-30.

Statistics Canada (2019). Statistics Canada Quality Guidelines. Technical Report 12-539-X. Accessed: 2020-08-24.

Statistics Canada (2020). Definitions, data sources and methods. Technical report, Statistics Canada. Accessed: 2020-07-30.

Statistics Canada (2021). National travel survey microdata file. Technical report, Statistics Canada. Accessed: 2022-07-18.

Statistics Canada (2022). North American industry classification system (NAICS) Canada 2022. Technical report, Statistics Canada. Accessed: 2023-03-08.

Suits, D. B. (1957). Use of dummy variables in regression equations. *Journal of the American Statistical Association*, 52(280):548–551.

Teucher, A., Albers, S., Hazlitt, S., and of British Columbia, P. (2023). *bcdata: Search and Retrieve Data from the BC Data Catalogue.* R package version 0.4.1.

Tukey, J. W. (1977). *Exploratory Data Analysis.* Addison-Wesley.

United Nations Statistics Division (2020). Statistical classifications. Technical report, United Nations. Accessed: 2020-09-07.

van der Loo, M. and de Jonge, E. (2018). *Statistical Data Cleaning with Applications in R.* Wiley.

van der Loo, M., de Jonge, E., and Hsieh, P. (2022). *validate: Data Validation Infrastructure.* R package version 1.1.1.

van der Loo, M. P. and de Jonge, E. (2021). Data validation infrastructure for R. *Journal of Statistical Software*, 97:1–33.

Vasilopoulos, K. (2022). *osnr: Client for the 'ONS' API.* R package version 1.0.1.

Venables, B. (2010). Introduction to Data Technologies. By Paul Murrell (review). *Australian and New Zealand Journal of Statistics*, 52:469–470.

von Bergmann, J. and Shkolnik, D. (2021). *cansim: Accessing Statistics Canada Data Table and Vectors.* R package version 0.3.14.

Walker, K. (2023). *Analyzing US Census Data: Methods, Maps, and Models in R.* CRC Press.

Walker, K., Herman, M., and Eberwein, K. (2023). *tidycensus: Load US Census Boundary and Attribute Data as 'tidyverse' and 'sf'-Ready Data Frames.* R package version 1.4.1.

Wang, R. Y., Reddy, M., and Kon, H. B. (1995). Toward quality data: An attribute-based approach. *Decision Support Systems*, 13(3):349–372.

Watson, J. (2020). *RBNZ: Download Data from the Reserve Bank of New Zealand Website.* R package version 1.1.0.

White, E., Baldridge, E., Brym, Z., and Locey, K. (2013). Nine simple ways to make it easier to (re)use your data. *Ideas in Ecology and Evolution*, 6(2):1–10.

Wickham, H. (2014). Tidy data. *Journal of Statistical Software*, 59(1):1–23.

Wickham, H. (2015). *Advanced R.* CRC Press.

Wickham, H. (2019a). *Advanced R.* CRC Press, 2 edition.

Wickham, H. (2019b). *stringr: Simple, Consistent Wrappers for Common String Operations.* R package version 1.4.0.

Wickham, H. (2021a). *forcats: Tools for Working with Categorical Variables (Factors).* R package version 0.5.1.

Wickham, H. (2021b). *tidyr: Tidy Messy Data.* R package version 1.1.3.

Wickham, H. (2023). *httr2: Perform HTTP Requests and Process the Responses.* https://httr2.r-lib.org, https://github.com/r-lib/httr2.

Wickham, H. and Bryan, J. (2019). *readxl: Read Excel Files.* R package version 1.3.1.

Wickham, H., Girlich, M., Ruiz, E., and RStudio (2023a). *dbplyr: A 'dplyr' Back End for Databases.* R package version 2.3.2.

Wickham, H. and Grolemund, G. (2016). *R for Data Science.* O'Reilly Media.

Wickham, H. and Hester, J. (2020). *readr: Read Rectangular Text Data.* R package version 1.4.0.

Wickham, H. and Miller, E. (2021). *haven: Import and Export SPSS, Stata and SAS Files.* R package version 2.4.1.

Wickham, H. and RStudio (2021). *nycflights13: Flights that Departed NYC in 2013.* R package version 1.0.2.

Wickham, H., Çetinkaya Rundel, M., and Grolemund, G. (2023b). *R for Data Science.* O'Reilly Media, 2 edition.

Wikipedia contributors (2021). Prawo jazdy (alleged criminal) — Wikipedia, the free encyclopedia. [Online; accessed 2022-01-10].

Wikipedia contributors (2022a). List of unicode characters — Wikipedia, the free encyclopedia. [Online; accessed 2022-10-03].

Wikipedia contributors (2022b). Pdf — Wikipedia, the free encyclopedia. [Online; accessed 2022-08-23].

Wilson, G., Aruliah, D. A., Brown, C. T., Hong, N. P. C., Davis, M., Guy, R. T., Haddock, S. H. D., Huff, K. D., Mitchell, I. M., Plumbley, M. D., Waugh, B., White, E. P., and Wilson, P. (2014). Best practices for scientific computing. *PLoS Biology*, 12(1).

Wilson, G., Bryan, J., Cranston, K., Kitzes, J., Nederbragt, L., and Teal, T. K. (2017). Good enough practices in scientific computing. *PLoS Computational Biology*, 13(6).

Xie, Y. (2021). *bookdown: Authoring Books and Technical Documents with R Markdown.* R package version 0.22.

Xie, Y., Allaire, J., and Grolemund, G. (2019). *R Markdown: the Definitive Guide.* CRC Press.

Zeeberg, B. R., Riss, J., Kane, D. W., Bussey, K. J., Uchio, E., Linehan, W. M., Barrett, J. C., and Weinstein, J. N. (2004). Mistaken identifiers: Gene name errors can be introduced inadvertently when using excel in bioinformatics. *BMC Bioinfomatics*, 5(80).

Zumel, N. and Mount, J. (2019). *Practical Data Science with R.* Manning, 2nd edition.

Index

Printed in the United States
by Baker & Taylor Publisher Services

Printed in the United States
by Baker & Taylor Publisher Services